I0038456

Wireless Communications and Networking: Concepts, Technologies and Applications

Wireless Communications and Networking: Concepts, Technologies and Applications

Stephen Morris

Larsen & Keller
www.larsen-keller.com

Wireless Communications and Networking: Concepts, Technologies and Applications
Stephen Morris
ISBN: 978-1-64172-661-0 (Hardback)

© 2022 Larsen & Keller

⊟ Larsen & Keller

Published by Larsen and Keller Education,
5 Penn Plaza,
19th Floor,
New York, NY 10001, USA

Cataloging-in-Publication Data

Wireless communications and networking : concepts, technologies and applications / Stephen Morris.
    p. cm.
Includes bibliographical references and index.
ISBN 978-1-64172-661-0
1. Wireless communication systems. 2. Wireless LANs. 3. Telecommunication systems.
I. Morris, Stephen.
TK5103.2 .W57 2022
621.384--dc23

This book contains information obtained from authentic and highly regarded sources. All chapters are published with permission under the Creative Commons Attribution Share Alike License or equivalent. A wide variety of references are listed. Permissions and sources are indicated; for detailed attributions, please refer to the permissions page. Reasonable efforts have been made to publish reliable data and information, but the authors, editors and publisher cannot assume any responsibility for the validity of all materials or the consequences of their use.

Trademark Notice: All trademarks used herein are the property of their respective owners. The use of any trademark in this text does not vest in the author or publisher any trademark ownership rights in such trademarks, nor does the use of such trademarks imply any affiliation with or endorsement of this book by such owners.

For more information regarding Larsen and Keller Education and its products, please visit the publisher's website www.larsen-keller.com

Table of Contents

Preface

The transfer of information or power between two or more points which are not connected by an electrical conductor is known as wireless communication. Most of the wireless technologies make use of radio waves. There are different devices which are used for wireless communication such as cellular telephones and two-way radios. Some of the other means of wireless communications are free space optical communication, sonic communication and electromagnetic induction. Wireless network refers to a network of computers where wireless data connections between network nodes are used. The topics included in this book on wireless communications are of utmost significance and bound to provide incredible insights to readers. Also included herein is a detailed explanation of the various concepts and applications of this field. This book will serve as a valuable source of reference for graduate and post graduate students.

A foreword of all Chapters of the book is provided below:

Chapter 1 - The collection of elements which function interdependently and which make use of unguided electromagnetic-wave propagation in order to perform specified functions is known as a wireless system. The topics elaborated in this chapter will help in gaining a better perspective about the wireless systems as well as wireless communications; **Chapter 2** - Analog communication makes use of a method of transmission where information is conveyed using a continuous signal which varies in phase, amplitude, or some other property with respect to that information. This chapter closely examines the key concepts of analog communication such as amplitude modulation, angle modulation and frequency modulation to provide an extensive understanding of the subject; **Chapter 3** - The mode of communication where information is encoded in a digital format and then transferred electronically is known as digital communication. This chapter discusses in detail the components and techniques related to digital communications and transmission such as sampling techniques, digital modulation techniques and M-ary encoding; **Chapter 4** - Wireless networking is a method through which different network nodes are connected through wireless data connections. Some of the different types of wireless networks are personal area networks, global area networks and cellular wireless networks. The topics elaborated in this chapter will help in gaining a better perspective about these types of wireless networks; **Chapter 5** - The group of spatially separated sensors which monitor and record the physical conditions of the environment and then organize the data in a central location is known as wireless sensor network. They can be used for health care monitoring and air pollution monitoring. The diverse aspects and applications of wireless sensor networks have been thoroughly discussed in this chapter.

I would like to thank the entire editorial team who made sincere efforts for this book and my family who supported me in my efforts of working on this book. I take this opportunity to thank all those who have been a guiding force throughout my life.

Stephen Morris

Wireless Systems: An Introduction

The collection of elements which function interdependently and which make use of unguided electromagnetic-wave propagation in order to perform specified functions is known as a wireless system. The topics elaborated in this chapter will help in gaining a better perspective about the wireless systems as well as wireless communications.

Wireless Communications

Wireless Communication is the fastest growing and most vibrant technological areas in the communication field. Wireless Communication is a method of transmitting information from one point to other, without using any connection like wires, cables or any physical medium.

Generally, in a communication system, information is transmitted from transmitter to receiver that is placed over a limited distance. With the help of Wireless Communication, the transmitter and receiver can be placed anywhere between few meters (like a T.V. Remote Control) to few thousand kilometer's (Satellite Communication).

We live in a World of communication and Wireless Communication, in particular is a key part of our lives. Some of the commonly used Wireless Communication Systems in our day-to-day life are: Mobile Phones, GPS Receivers, Remote Controls, Bluetooth Audio and Wi-Fi etc.

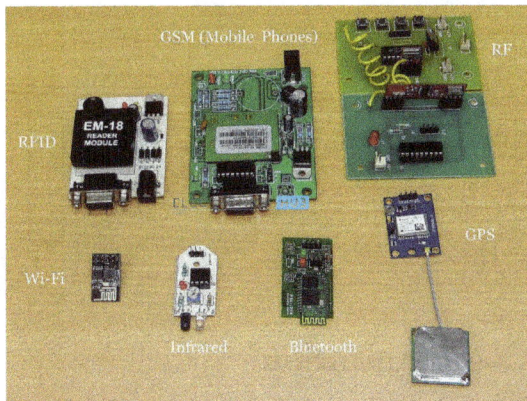

Communication systems can be wired or wireless and the medium used for communication can be guided or unguided. In wired communication, the medium is a physical path like co-axial cables, twisted pair cables and optical fiber links etc. which guides the signal to propagate from one point to other.

Such type of medium is called guided medium. On the other hand, wireless communication doesn't require any physical medium but propagates the signal through space. Since, space only allows for signal transmission without any guidance, the medium used in wireless communication is called unguided medium.

If there is no physical medium, then how does wireless communication transmit signals? Even though there are no cables used in wireless communication, the transmission and reception of signals is accomplished with Antennas.

Antennas are electrical devices that transform the electrical signals to radio signals in the form of Electromagnetic (EM) Waves and vice versa. These Electromagnetic Waves propagates through space. Hence, both transmitter and receiver consist of an antenna.

Electromagnetic Wave

Electromagnetic Waves carry the electromagnetic energy of electromagnetic field through space. Electromagnetic Waves include Gamma Rays (γ-Rays), X-Rays, Ultraviolet Rays, Visible Light, Infrared Rays, Microwave Rays and Radio Waves. Electromagnetic Waves (usually Radio Waves) are used in wireless communication to carry the signals.

An Electromagnetic Wave consists of both electric and magnetic fields in the form of time varying sinusoidal waves. Both these fields are oscillating perpendicular to each other and the direction of propagation of the Electromagnetic Wave is again perpendicular to both these fields.

Mathematically, an Electromagnetic Wave can be described using Maxwell's equations. Pictorial representation of an Electromagnetic Wave is shown below where the Electric Field is acting in the Y-axis, magnetic field is acting in the Z-axis and the Electromagnetic Wave propagates in X-axis.

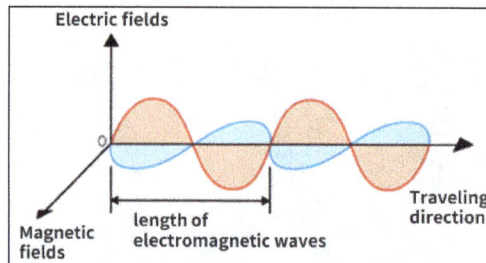

Need for Wireless Communication

When wired communication can do most of the tasks that a wireless communication can, why do we need Wireless Communication? The primary and important benefit of wireless communication is mobility.

Apart from mobility, wireless communication also offers flexibility and ease of use, which makes it increasingly popular day-by-day. Wireless Communication like mobile telephony can be made anywhere and anytime with a considerably high throughput performance.

Another important point is infrastructure. The setup and installation of infrastructure for wired communication systems is an expensive and time consuming job. The infrastructure for wireless communication can be installed easily and low cost.

In emergency situations and remote locations, where the setup of wired communication is difficult, wireless communication is a viable option.

Advantages of Wireless Communication

There are numerous advantages of Wireless Communication Technology, Wireless Networking and Wireless Systems over Wired Communication like Cost, Mobility, Ease of Installation, and Reliability etc.

Cost

The cost of installing wires, cables and other infrastructure is eliminated in wireless communication and hence lowering the overall cost of the system compared to wired communication system. Installing wired network in building, digging up the Earth to lay the cables and running those wires across the streets is extremely difficult, costly and time consuming job.

In historical buildings, drilling holes for cables is not a best idea as it destroys the integrity and importance of the building. Also, in older buildings with no dedicated lines for communication, wireless communication like Wi-Fi or Wireless LAN is the only option.

Mobility

Mobility is the main advantage of wireless communication system. It offers the freedom to move around while still connected to network.

Ease of Installation

The setup and installation of wireless communication network's equipment and infrastructure is very easy as we need not worry about the hassle of cables. Also, the time required to setup a wireless system like a Wi-Fi network for example, is very less when compared to setting up a full cabled network.

Reliability

Since there are no cables and wires involved in wireless communication, there is no chance of communication failure due to damage of these cables which may be caused by environmental conditions, cable splice and natural diminution of metallic conductors.

Disaster Recovery

In case of accidents due to fire, floods or other disasters, the loss of communication infrastructure in wireless communication system can be minimal.

Disadvantages of Wireless Communication

Even though wireless communication has a number of advantages over wired communication, there are a few disadvantages as well. The most concerning disadvantages are Interference, Security and Health.

Interference

Wireless Communication systems use open space as the medium for transmitting signals. As a result, there is a huge chance that radio signals from one wireless communication system or network might interfere with other signals.

The best example is Bluetooth and Wi-Fi (WLAN). Both these technologies use the 2.4 GHz frequency for communication and when both of these devices are active at the same time, there is a chance of interference.

Security

One of the main concerns of wireless communication is Security of the data. Since the signals are transmitted in open space, it is possible that an intruder can intercept the signals and copy sensitive information.

Health Concerns

Continuous exposure to any type of radiation can be hazardous. Even though the levels of RF energy that can cause the damage are not accurately established, it is advised to avoid RF radiation to the maximum.

Modes of Communications

Microwave Transmission

Microwave radio, a form of radio transmission that use. Ultra-high frequencies developed out of experiments with radar (radio detecting and ranging) during the period preceding World War II. There are several frequency ranges assigned to microwave systems, all of which are in the Giga Hertz (GHz) range and the wavelength in the millimeter range. This very short wavelength gives rise to the term microwave. Such high frequency signals are especially susceptible to attenuation and, therefore must be amplified or repeated after a particular distance.

In order to maximize the strength of such a high frequency signal and, therefore, to increase the distance of transmission at acceptable levels, the radio beams are highly focused. The transmit antenna is centered in a concave, reflective metal dish which serves to focus the radio beam with maximum effect on the receiving antenna, as illustrated in figure. The receiving antenna, similarly, is centered in a concave metal dish, which serves to collect the maximum amount of incoming signal.

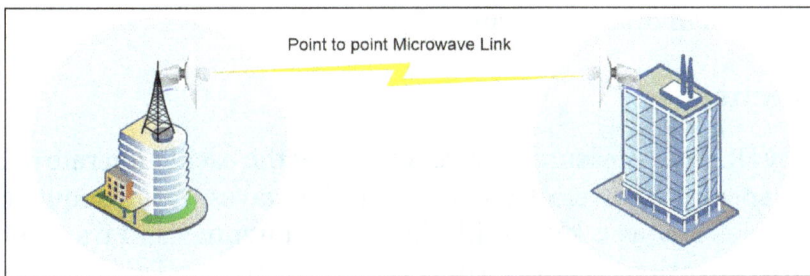

Point to point Microwave Link

It is a point-to-point, rather than a broadcast, transmission system. Additionally, each antenna must be within line of sight of the next antenna. Given the curvature of the earth, and the obvious problems of transmitting through it, microwave hops generally are limited to 50 miles (80 km). If the frequencies are higher within the microwave band, this impact is more than lower frequencies in the same band.

Frequency bands maximum antenna separation analog/digital 4-6 GHz 32-48 km, Analog 10-12 GHz 16-24 km, Digital 18-23 GHz 8-11 km.

Properties of Microwave Transmission

- Configuration - Microwave radio consists of antennae centered within reflective dishes that are attached to structures such as towers or buildings. Cables connect the antennae to the actual transmit (receive) equipment.

- Bandwidth - Microwave offers substantial bandwidth, often in excess of 6 Gbps.

- Error Performance - Microwave, especially digital microwave, performs well in this regard, assuming proper design. However, such high frequency radio is particularly susceptible to environmental interference, e.g. precipitation, haze, smog, and smoke. Generally speaking, however, microwave performs well in this regard.

- Distance Microwave - Clearly is distance limited, especially at the higher frequencies. This limitation can be mitigated through special and more complex arrays of antennae incorporating spatial diversity in order to collect more signals.

- Security - As is the case with all radio systems, microwave is inherently not secure. Security must be imposed through encryption (scrambling) of the signal.

- Cost - The acquisition, deployment and rearrangement cost of microwave can be high. However, it often compares very favorably with cabled systems, which require right-of-way, trenching, conduit, splicing, etc.

- Applications - Microwave originally was used for long haul voice and data communications. Competing long distance carriers, microwave was found the most attractive alternative to cabled systems, due to the speed and low cost qf deployment where feasible, however, fiber optic technology is currently used in this regard. Contemporary applications include private networks, interconnection of cellular radio switches, and as an alternative to cabled systems in consideration of difficult terrain.

Radio Frequency

Radio frequency (RF) is a measurement representing the oscillation rate of electromagnetic radiation spectrum, or electromagnetic radio waves, from frequencies ranging from 300 GHz to as low as 9 kHz. With the use of antennas and transmitters, an RF field can be used for various types of wireless broadcasting and communications.

How Radio Frequency Works

Radio frequency is measured in units called hertz, which represent the number of cycles per second when a radio wave is transmitted. One hertz equals one cycle per second; radio waves range from thousands (kilohertz) to millions (megahertz) to billions (gigahertz) of cycles per second. Microwaves are a type of radio wave with higher frequencies. Radio frequencies are not visible to the human eye.

In a radio wave, the wavelength is inversely proportional to the frequency. If f is the frequency in megahertz and s is the wavelength in meters, then s = 300/f.

As the frequency is increased beyond that of the RF spectrum, electromagnetic energy takes the form of infrared (IR), visible, ultraviolet, X-rays and gamma rays.

RF Technology

Many types of wireless devices make use of RF fields. Cordless and cellphones, radio and television broadcast stations, Wi-Fi and Bluetooth, satellite communications systems, and two-way radios all operate in the RF spectrum. In addition, other appliances outside of communications, including microwave ovens and garage-door openers, operate at radio frequencies. Some wireless devices, like TV remote controls, some cordless computer keyboards and computer mice, operate at IR frequencies, which have shorter electromagnetic wavelengths.

The RF spectrum is divided into several ranges, or bands. With the exception of the lowest-frequency segment, each band represents an increase of frequency corresponding to an order of magnitude (power of 10). The following table depicts the eight bands in the RF spectrum, showing frequency and bandwidth ranges. The super high frequency (SHF) and extremely high frequency (EHF) bands are often referred to as the *microwave spectrum*.

Radio frequency spectrum bands			
Designation	Abbrevition	Frequencies	Free-space wavelengths
Very low frequency	VLF	9 KHz to 30 KHz	33 km to 10 km
Low frequency	LF	30 KHz to 300 KHz	10 km to 1 km
Medium frequency	MF	300 KHz to 3 MHz	1 km to 100 m
High frequency	HF	3 MHz to 300 MHz	100 m to 10 m
Very high frequency	VHF	30 MHz to 300 MHz	10 m to 1 m
Ultrahigh frequency	UHF	300 MHz to 3 GHz	1 m to 100 mm
Super high frequency	SHF	3 GHz to 30 GHz	100 mm to 10 mm
Extremely frequency	EHF	30 GHz to 300 GHz	10 mm to 1 mm

In the United States, radio frequencies are divided into licensed and unlicensed bands. The Federal Communications Commission issues licenses that permit a commercial entity to have exclusive use of a frequency band in a given location. Unlicensed frequencies are free for public use, but remain a shared medium.

Free Space Optics

Free Space Optics (FSO) communications refers to the transmission of modulated visible or infrared beams through the atmosphere to obtain optical communications. FSO demonstrates a typical free space optical link operating at 1.25 GB/s, where usually the main source of penalty is the atmospheric attenuation.

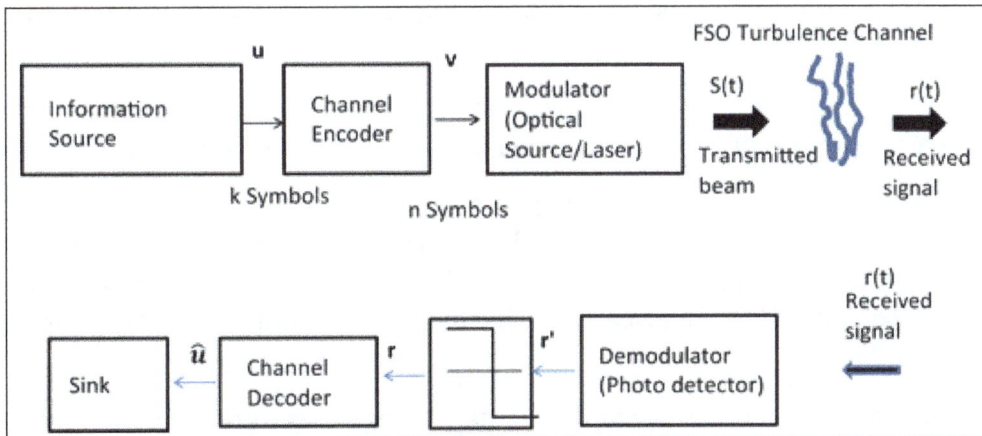

The wireless optical channel component, that is also free-space optics, can be used for large distances where the atmospheric attenuation is not the major source of penalties, but the pointing angle is. e.g. satellite communications. WOC demonstrates the impaired link performance from pointing errors between the transmitter and receiver for satellite communications.

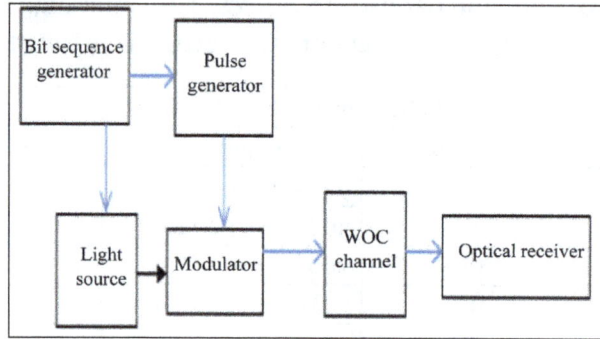

WOC.

Applications

Vehicles

Many wireless communication systems and mobility aware applications are used for following purpose:

- Transmission of music, news, road conditions, weather reports, and other broadcast information are received via digital audio broadcasting (DAB) with 1.5 Mbit/s.

- For personal communication, a universal mobile telecommunications system (UMTS) phone might be available offering voice and data connectivity with 384 kbit/s.

- For remote areas, satellite communication can be used, while the current position of the car is determined via the GPS (Global Positioning System).

- A local ad-hoc network for the fast exchange of information (information such as distance between two vehicles, traffic information, road conditions) in emergency situations or to help each other keep a safe distance. Local ad-hoc network with vehicles close by to prevent guidance system, accidents, redundancy.

- Vehicle data from buses, trucks, trains and high speed train can be transmitted in advance for maintenance.

- In ad-hoc network, car can comprise personal digital assistants (PDA), laptops, or mobile phones connected with each other using the Bluetooth technology.

A typical application of mobile communication in road traffic.

Emergency

Following services can be provided during emergencies:

- Video communication: Responders often need to share vital information. The transmission of real time situations of video could be necessary. A typical scenario includes the transmission of live video footage from a disaster area to the nearest fire department, to the police station or to the near NGOs etc.

- Push To Talk (PTT): PTT is a technology which allows half duplex communication between two users where switching from voice reception mode to the transmit mode takes place with the use of a dedicated momentary button. It is similar to walkie-talkie.

- Audio/Voice Communication: This communication service provides full duplex audio channels unlike PTT. Public safety communication requires novel full duplex speech transmission services for emergency response.

- Real Time Text Messaging (RTT): Text messaging (RTT) is an effective and quick solution for sending alerts in case of emergencies. Types of text messaging can be email, SMS and instant message.

Business

Travelling Salesman

- Directly access to customer files stored in a central location.
- Consistent databases for all agents.
- Mobile office.
- To enable the company to keep track of all the activities of their travelling employees.

In Office

- Wi-Fi wireless technology saves businesses or companies a considerable amount of money on installations costs.

- There is no need to physically setup wires throughout an office building, warehouse or store.

- Bluetooth is also a wireless technology especially used for short range that acts as a complement to Wi-Fi. It is used to transfer data between computers or cellphones.

Transportation Industries

In transportation industries, GPS technology is used to find efficient routes and tracking vehicles.

Replacement of Wired Network

- Wireless network can also be used to replace wired network. Due to economic reasons it is often impossible to wire remote sensors for weather forecasts, earthquake detection, or to provide environmental information, wireless connections via satellite, can help in this situation.

- Tradeshows need a highly dynamic infrastructure, since cabling takes a long time and frequently proves to be too inflexible.

- Many computers fairs use WLANs as a replacement for cabling.

- Other cases for wireless networks are computers, sensors, or information displays in historical buildings, where excess cabling may destroy valuable walls or floors.

Location Dependent Service

It is important for an application to know something about the location because the user might need location information for further activities. Several services that might depend on the actual location can be described below:

Follow-on Services

- Location aware services: To know about what services (e.g. fax, printer, server, phone, printer etc.) exist in the local environment.

- Privacy: We can set the privacy like who should get knowledge about the location.

- Information Services: We can know about the special offers in the supermarket. Nearest hotel, rooms, cabs etc.

Infotainment

- Wireless networks can provide information at any appropriate location.

- Outdoor internet access.

- You may choose a seat for movie, pay via electronic cash, and send this information to a service provider.

- Ad-hoc network is used for multiuser games and entertainment.

Mobile and Wireless Devices

Even though many mobile and wireless devices are available, there will be many more devices in the future. There is no precise classification of such devices, by sizes, shape, weight, or computing power. The following list of given examples of mobile and wireless devices graded by increasing performance (CPU, memory, display, input devices, etc).

- Sensor: Wireless device is represented by a sensor transmitting state information. One example could be a switch, sensing the office door. If the door is closed, the switch transmits this information to the mobile phone inside the office which will not accept incoming calls without user interaction; the semantics of a closed door is applied to phone calls.

- Embedded Controller: Many applications already contain a simple or sometimes more complex controller. Keyboards, mouse, headsets, washing machines, coffee machines, hair dryers and TV sets are just some examples.

- Pager: As a very simple receiver, a pager can only display short text messages, has a tiny display, and cannot send any messages.

- Personal Digital Assistant: PDAs typically accompany a user and offer simple versions of office software (calendar, notepad, mail). The typically input device is a pen, with built-in character recognition translating handwriting into characters. Web browsers and many other packages are available for these devices.

- Pocket computer: The next steps towards full computers are pocket computers offering tiny keyboards, color displays, and simple versions of programs found on desktop computers (text processing, spreadsheets etc).

- Notebook/laptop: Laptops offer more or less the same performance as standard desktop computers; they use the same software - the only technical difference being size, weight, and the ability to run on a battery. If operated mainly via a sensitive display (touch sensitive or electromagnetic), the device are also known as notepads or tablet PCs.

Components of Communication Systems

The communication system basically deals with the transmission of information from one point to another using the well-defined steps which are carried out in sequential manner. The system for data transmission makes use of the sender and destination address in this, other elements are also there that allows it to transfer data from one set of point to another set of point after dividing the elements of communication system in groups and these interface elements acts as the main component for data communication and all these interface elements are given below.

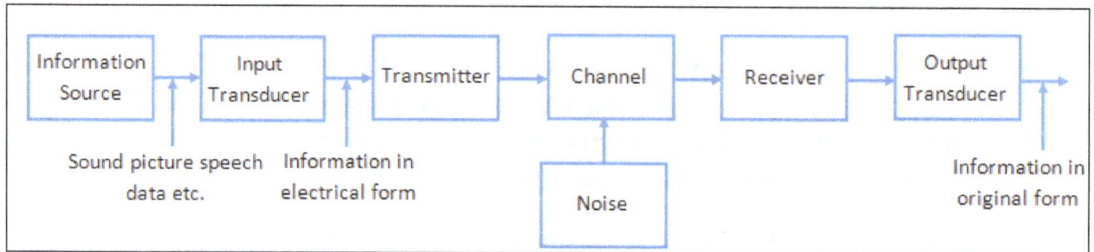

Information Source

The communication system which we are using is act as the main communication source for data transmission between two machines. Firstly, the source of data code is generated either in numeric form or in character form such that it should be in encrypted manner that does not provide information access to unknown or unauthorized user, this unit uses the specialized tools and utilities for the generation of messages which is to be transmitted over the communication channel such that the signal can either be analog or digital in nature and it is converted from one form to another according to the compatibility of transmission medium that represents the signal nature. Moreover, the data source which is generated using the encoder has filter component that refines the data packets and removes data redundancy using the normalization technique.

Input Transducer

As you know that basic work of the transducer is to convert one form of energy into another form that can be electrical in nature. Let us consider that input source signal is non- electrical in nature then you have to first convert these signals in time varying electrical signal. For example: the microphone which we use in seminars and presentations converts message information into sound waves which is electrical in nature. Once you have successfully converted it into electrical signals then data compression technique is used which will compress data packets into single package so that it can be easily transmitted over the transmission lines because data compression reduces the size of the data packets to be transmitted.

Transmitter

The source generated electrical signals are then used by the transmitter after refining them and removes the noise and distortion there in it and makes signal in form that can be easily amplified, for the purpose of amplification in transmitter circuit we uses the digital modulator that converts sequence into electrical signals so that it can be easily transmitted over long distance. For example, in the wire telephony system, the modulation is used for the enhancement of the signal strength without the loss of the original data because using the ordinary antenna's it is not possible to reduce the noise and distortion during transmission of data signal.

Communication Channel

The physical medium which is used for the transmission of communication data signals from sender to receiver is referred as communication channel and we can also say that it is the platform that allows the sending and receiving of the data packets using the well-established path between two machines that can either be wire oriented or wireless such that both types of connections are supported by the point to point and broadcast channel, the various communication channels are used in it for the data transmission that depends on the type of the network topology and circuit which we are using. Instead of this, the optical media is the best communication channel that provides fast and safe data transmission because tracing of the signals in it is impossible.

Receiver

The receiver machine work is to reproduce the message signal in electrical form from the noised and distorted signal such that digital demodulator is used that process the waveform signals into the sequence of numbers that represents the discrete values which is in form of zeros and ones and then these discrete signals are used for the reconstruction of information code from the attenuated signal.

Destination Machine

The last stage of the communication system is destination machine which converts these electrical signals into its original form for the data broadcasting so that it can be easily understand by the end user or receiver and then this same sort of communication process is used for the acknowledgment of signals to sender machine.

References

- Wireless-communication-introduction-types-applications: electronicshub.org, Retrieved 4 March, 2019

- Microwave-transmission, communication-networks: ecomputernotes.com, Retrieved 14 May, 2019

- Radio-frequency: techtarget.com, Retrieved 24 January, 2019

- Optical-system-free-space-optics-fso: optiwave.com, Retrieved 2 June, 2019

- Applications-of-wireless-communication: javatpoint.com, Retrieved 17 April, 2019

- 4509-Basic-Elements-used-Communication-System: techulator.com, Retrieved 7 February, 2019

2

Analog Communication

Analog communication makes use of a method of transmission where information is conveyed using a continuous signal which varies in phase, amplitude, or some other property with respect to that information. This chapter closely examines the key concepts of analog communication such as amplitude modulation, angle modulation and frequency modulation to provide an extensive understanding of the subject.

Analog communication is a data transmitting technique in a format that utilizes continuous signals to transmit data including voice, image, video, electrons etc. An analog signal is a variable signal continuous in both time and amplitude which is generally carried by use of modulation. Along with the working in analog communication it is used in various technique.

A communication system is made up of devices that employ one of two communication methods (wireless or wired), different types of equipment (portable radios, mobile radios, base/fixed station radios, and repeaters), and various accessories (examples include speaker microphones, battery eliminators, and carrying cases) and enhancements (encryption, digital communications, security measures, and interoperability/ networking) to meet the user needs.

A communication system can be considered to be "wired" or "wireless" (e.g., conventional telephone, radio communications, etc.) A wired system is technically known as a hard-line system and can be thought of as a localized, private telephone system that uses wires to operate over a limited area. A wireless system uses radio frequencies to "connect" users and is capable of operating over a much larger geographical area than a hard-line (wired) system.

Hard-line Technology

Hard-line communication systems operate by transmitting voice and data through a cable that connects to a telephone-like apparatus. The major advantage of a hard-line system is the ability to communicate from underground, confined spaces, shielded enclosures, collapsed structure void spaces, and similar locations (such as explosive environments) where RF systems are unreliable or unable to be used.

Since the communication equipment available to emergency first responders today does not use optical transmission methods only radio frequency (RF) equipment will be considered here. Shared communication systems such as radios, the Internet, and telephone conference calls are subject to saturation by users (the maximum capacity whereby adding users will deteriorate and degrade the amount and quality of information able to be transferred over the system), a problem that compounds exponentially as the number of users increases. Communication system efficiency requires that the users follow published communication system guidelines regarding proper system discipline in order to ensure maximum efficiency of communication traffic.

A radio apparatus has a section for receiving an analog signal and a digital angle-modulated carrier wave signal. The analog signal and digital angle-modulated carrier wave output of the receiver section are demodulated to provide first and second demodulated signals. A clock signal is regenerated from the output of either the demodulating means or receiver section. A control signal selectively operates a switch for passing either the first or the second demodulated signals. The regenerated clock signal controls the switch.

The radio equipment involved in communication systems includes a transmitter and a receiver, each having an antenna and appropriate terminal equipment such as a microphone at the transmitter and a loudspeaker at the receiver in the case of a voice-communication system.

A transmitter is an electronic device which, usually with the aid of an antenna, propagates an electromagnetic signal such as radio, television, or other telecommunications.

The analog communication system uses a modified version of the high sensitivity homodyne Syncbit data transmission principle and uses the proven Nd:YAG laser technology operating at 1064 nm. Therefore, it can use the same electro-optical building blocks (lasers, modulator, laser amplifier) as the digital system.

The analog communication system fits into the modular concept of the OPTEL terminals and is fully compatible to all optical heads of the terminal family.

Use in Every Day Life

Portable Radios

Portable radios are small, lightweight, handheld, wireless communication units that contain both a transmitter and a receiver, a self-contained microphone and speaker, an attached power supply (typically a rechargeable battery), and antenna. Portable transceivers (such as a walkie-talkie) have relatively low-powered transmitters (1 W to 5 W), need to have their batteries periodically recharged or replaced, and may be combined in a wireless radio communication system with other portable, mobile, and base station radios. There are also very low-powered transceivers, available with power outputs of 0.1 W, which are generally linked to portable repeaters for extended range and interoperability with higher-powered radio systems.

Mobile Radios

Mobile radios are larger than portable radios and are designed to be mounted in a fixed location inside a vehicle (police cruiser, fire truck, etc.). Like the portable radios, mobile radios contain both a transmitter and a receiver and may contain an internal speaker. However, mobile radios connect to the vehicle's power supply, which enables them to have a higher transmitter output power (typically 5 W to 50 W) and an external antenna. The microphone is usually handheld, and the speaker may be externally located to the radio. Because of the higher transmitter power and external antenna, the effective communication range is greater than that of a portable radio, especially if a repeater is not used.

Repeaters

A repeater is a specialized radio that contains both a receiver and a transmitter. Repeaters are used to increase the effective communications coverage area for portable, mobile, or base station radios that otherwise might not be able to communicate with one another. The repeater's receiver is tuned to the frequency used by a portable, mobile, or base station transmitter for incoming signals, and the repeater's transmitter is tuned to the frequency used by a portable, mobile, or base station receiver.

In Lab Practicals

Our study contains both computer-simulated applications and real time applications. The front panels of the real time applications have been presented here. The front panels of real time applications of AM, PCM, ASK, FSK and PSK which are designed by using Lab view program.

Such a system provides a high speech quality. Well known among these digital angle modulation systems are phase shift keying (PSK) and frequency shift keying (FSK) systems. The PSK modulation system is an excellent system for the transmission of a large quantity of information per unit of frequency band and for requiring no large signal-to-noise power ratio in order to reduce the error rate to a given level. The FSK modulation system can cause class C amplifiers, or the like, to act nonlinearly because such an amplifier has a constant amplitude component. It is also superior in power efficiency and can be effectively used for apparatus with small battery capacities, such as mobile communication units.

The front panels called "user interfaces" are designed as authentically as possible. The front panels of these applications that contain two sections such as the control buttons on the left and input-output waveforms of the experiments on the right are designed in a simple and intelligible structure.

The frequency of sample impulses (PCM), carrier frequencies (ASK, FSK, PSK, AM) are changed as virtual and the frequency of message signals are changed as real time by using control buttons. So students can change the message signals in real time via voltage controlled circuit (VCO) by controlling the connected devices. The measured results from the instruments are brought to the user through the same interface.

There are five types of signals such as sinusoidal, square, triangle, saw tooth and random on the message signal part. In addition, the front panels display message signal, sample impulses, carrier signals as input and AM modulated signal, PCM Code Sequence, ASK, FSK, PSK Signals as output.

Interoperability and Networking

Interoperability is the process of connecting different groups using different radio systems and communication technologies (telephones, radios, cellular communications, and satellite Communications) so that they can communicate directly with one another without having to go through multiple dispatchers or relay personnel. In the context of communications, interoperability describes the situation where different communication systems that are otherwise incompatible with one another work together without relying on the addition of considerably more manpower.

Signal path block diagram for an analog crosslink transceiver showing the building blocks of:

- Analog communication system (green).

- Laser subsystem (magenta).

- General communication functions, common to analog and to digital comms system (blue).

They also shows for information the building blocks of a digital communication system as outlined boxes and the 3 switches to select either analog or digital operation.

Analog in Hardware System

The hardware system and its components of the real time Analog and Digital Communication experiments. In this structure, Laboratory PC which includes GPIB and Ethernet Interfaces is called as a server PC. It works as the main controller. By using Voltage Controlled Oscillator (VCO) a message signal has been obtained for experiments. The input signal of VCO is controlled by the DC power supply with GPIB and the output signal of VCO circuit is measured with digital multimeter and then sent back to the Laboratory PC through the GPIB. This measured signal is applied to the system as a message frequency.

Analog Communication in Antenna

In the design of the analog communication system a precondition was to use the same electro-optic building blocks in the analog as in the digital terminals. The analog communication system module is another extension of our modular product family and fits together with all other modules of our terminals.

Crosslink Concept

In an analog satellite user data is contained in individual channels. Each channel has a certain bandwidth, for instance 36 MHz, and certain distance from the neighboring channels. In most cases channels are lined up in an equidistant frequency raster, the channel spacing, for instance, 41 MHz. The analog crosslink terminal accepts all channels from the uplink to satellite "A" which have to be transmitted to the counter terminal on satellite "B" and modulates them onto the optical carrier.

The laser beam is transmitted through the free-space channel to satellite "B". The optical counter terminal on satellite "B" receives the optical carrier, extracts the channels and inserts them into the microwave communications payload of satellite "B". Satellite "B" can now distribute the information in the usual way.

Crosslink Concept with Baseline Crosslink

- Channel rearrangement.

- Parallel distribution by on board information duplication.

For different overall signal to noise ratio or bit error rate requirements the ISL SNR can be easily increased or relaxed by varying the optical output power of the terminal transmitter.

Analog Communication in Speech Communication

Although speakers can use analog variation of acoustic properties to express the same information conveyed by lexical items in a sentence, can analog expression convey information that is independent of the propositional content or does it function just to modulate propositional meaning? If analog acoustic expression is a channel of communication that is truly different from the linguistic propositional channel, speakers should be able to express information that is different from the information conveyed in words and sentences. Experiment tested whether speakers can indeed use this channel to express information that is independent of the propositional content of the utterance and importantly, to see whether listeners are sensitive to the message conveyed exclusively through analog acoustic expression. If analog acoustic expression serves a communicative function, listeners should be able to understand the information it conveys, even when this information is not conveyed by any of the lexical items used by speakers.

Analog acoustic expression serves a communicative function by providing listeners with a "channel" of information over and above the propositional-linguistic content of the utterance. Furthermore, analog acoustic expression may facilitate comprehension by setting up a non-arbitrary mapping between form and meaning adding to the information provided by the arbitrary form-meaning mapping in the linguistic channel. Although in some cases analog acoustic expression may result from the speaker's communicative intention to convey a specific message through acoustic variation (the baseball example given earlier may be such an example), the results here are consistent with the idea that analog acoustic expression does not require a specific communicative intention on the part of the speaker.

Modulation

For a signal to be transmitted to a distance, without the effect of any external interference or noise addition and without getting faded away, it has to undergo a process called as Modulation. It improves the strength of the signal without disturbing the parameters of the original signal.

A message carrying a signal has to get transmitted over a distance and for it to establish a reliable communication; it needs to take the help of a high frequency signal which should not affect the original characteristics of the message signal.

The characteristics of the message signal, if changed, the message contained in it also alters. Hence, it is a must to take care of the message signal. A high frequency signal can travel up to a longer distance, without getting affected by external disturbances. We take the help of such high frequency signal which is called as a carrier signal to transmit our message signal. Such a process is simply called as Modulation.

Modulation is the process of changing the parameters of the carrier signal, in accordance with the instantaneous values of the modulating signal.

Need for Modulation

Baseband signals are incompatible for direct transmission. For such a signal, to travel longer distances, its strength has to be increased by modulating with a high frequency carrier wave, which doesn't affect the parameters of the modulating signal.

Advantages of Modulation

The antenna used for transmission, had to be very large, if modulation was not introduced. The range of communication gets limited as the wave cannot travel a distance without getting distorted.

Following are some of the advantages for implementing modulation in the communication systems:

- Reduction of antenna size.
- No signal mixing.
- Increased communication range.

- Multiplexing of signals.

- Possibility of bandwidth adjustments.

- Improved reception quality.

Signals in the Modulation Process

Following are the three types of signals in the modulation process:

Message or Modulating Signal

The signal which contains a message to be transmitted, is called as a message signal. It is a baseband signal, which has to undergo the process of modulation, to get transmitted. Hence, it is also called as the modulating signal.

Carrier Signal

The high frequency signal, which has a certain amplitude, frequency and phase but contains no information is called as a carrier signal. It is an empty signal and is used to carry the signal to the receiver after modulation.

Modulated Signal

The resultant signal after the process of modulation is called as a modulated signal. This signal is a combination of modulating signal and carrier signal.

Types of Modulation

There are many types of modulations. Depending upon the modulation techniques used, they are classified as shown in the following figure.

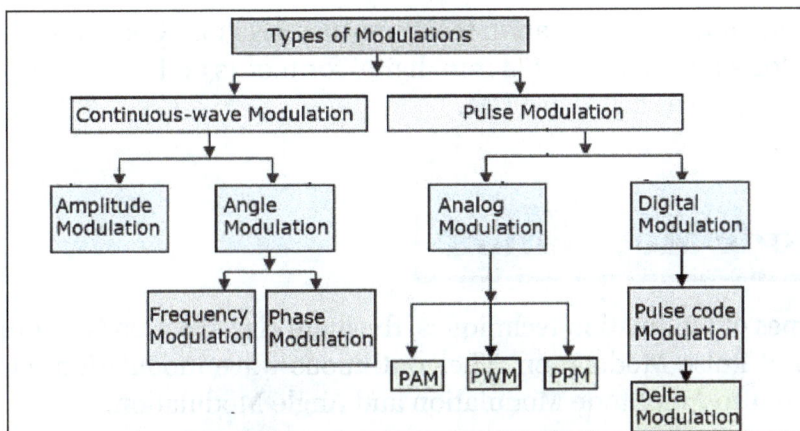

The types of modulations are broadly classified into
continuous-wave modulation and pulse modulation.

Continuous-wave Modulation

In continuous-wave modulation, a high frequency sine wave is used as a carrier wave. This is further divided into amplitude and angle modulation.

- If the amplitude of the high frequency carrier wave is varied in accordance with the instantaneous amplitude of the modulating signal, then such a technique is called as Amplitude Modulation.

- If the angle of the carrier wave is varied, in accordance with the instantaneous value of the modulating signal, then such a technique is called as Angle Modulation. Angle modulation is further divided into frequency modulation and phase modulation.

    ◦ If the frequency of the carrier wave is varied, in accordance with the instantaneous value of the modulating signal, then such a technique is called as Frequency Modulation.

    ◦ If the phase of the high frequency carrier wave is varied in accordance with the instantaneous value of the modulating signal, then such a technique is called as Phase Modulation.

Pulse Modulation

In Pulse modulation, a periodic sequence of rectangular pulses, is used as a carrier wave. This is further divided into analog and digital modulation.

In analog modulation technique, if the amplitude or duration or position of a pulse is varied in accordance with the instantaneous values of the baseband modulating signal, then such a technique is called as Pulse Amplitude Modulation (PAM) or Pulse Duration/Width Modulation (PDM/PWM), or Pulse Position Modulation (PPM).

In digital modulation, the modulation technique used is Pulse Code Modulation (PCM) where the analog signal is converted into digital form of 1s and 0s. As the resultant is a coded pulse train, this is called as PCM.

Amplitude Modulation

Among the types of modulation techniques, the main classification is Continuous-wave Modulation and Pulse Modulation. The continuous wave modulation techniques are further divided into Amplitude Modulation and Angle Modulation.

A continuous-wave goes on continuously without any intervals and it is the baseband message signal, which contains the information. This wave has to be modulated.

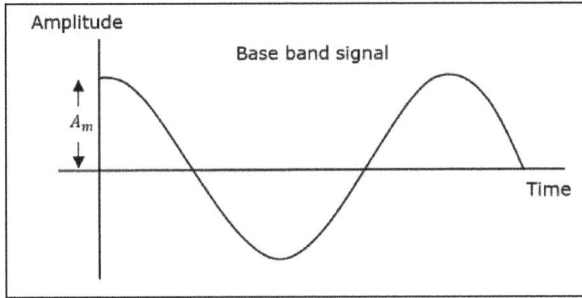

According to the standard definition, "The amplitude of the carrier signal varies in accordance with the instantaneous amplitude of the modulating signal". Which means, the amplitude of the carrier signal which contains no information varies as per the amplitude of the signal, at each instant, which contains information? This can be well explained by the following figures.

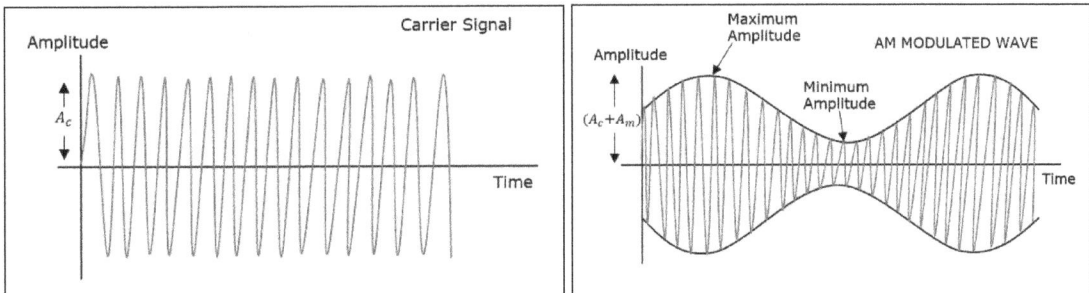

The modulating wave which is shown first is the message signal. The next one is the carrier wave, which is just a high frequency signal and contains no information. While the last one is the resultant modulated wave.

It can be observed that the positive and negative peaks of the carrier wave are interconnected with an imaginary line. This line helps recreating the exact shape of the modulating signal. This imaginary line on the carrier wave is called as Envelope. It is the same as the message signal.

Mathematical Expression

Following are the mathematical expression for these waves:

Time-domain Representation of the Waves.

Let modulating signal be:

$$m(t) = A_m \cos(2\pi f_m t)$$

Let carrier signal be:

$$c(t) = A_c \cos(2\pi f_c t)$$

Where, A_m = maximum amplitude of the modulating signal.

A_c = maximum amplitude of the carrier signal.

The standard form of an Amplitude Modulated wave is defined as:

$$S(t) = A_c \left[1 + K_a m(t)\right] \cos\left(2\pi f_c t\right)$$

$$S(t) = A_c \left[1 + \mu \cos\left(2\pi f_m t\right)\right] \cos\left(2\pi f_c t\right)$$

Where, $\mu = K_a A_m$.

Modulation Index

A carrier wave, after being modulated, if the modulated level is calculated, then such an attempt is called as Modulation Index or Modulation Depth. It states the level of modulation that a carrier wave undergoes.

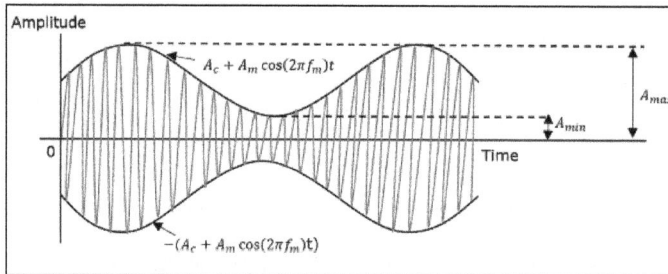

The maximum and minimum values of the envelope of the modulated wave are represented by A_{max} and A_{min} respectively.

Let us try to develop an equation for the Modulation Index.

$$A_{max} = A_c \left(1 + \mu\right)$$

Since, at A_{max} the value of cos θ is 1:

$$A_{min} = A_c \left(1 - \mu\right)$$

Since, at A_{min} the value of cos θ is -1:

$$\frac{A_{max}}{A_{min}} = \frac{1 + \mu}{1 - \mu}$$

$$A_{max} - \mu A_{max} = A_{min} + \mu A_{min}$$

$$-\mu\left(A_{max} + A_{min}\right) = A_{min} - A_{max}$$

$$\mu = \frac{A_{max} - A_{min}}{A_{max} + A_{min}}$$

Hence, the equation for Modulation Index is obtained. μ denotes the modulation index or modulation depth. This is often denoted in percentage called as percentage modulation. It is the extent of modulation denoted in percentage, and is denoted by m.

For a perfect modulation, the value of modulation index should be 1, which means the modulation depth should be 100%.

For instance, if this value is less than 1, i.e., the modulation index is 0.5, then the modulated output would look like the following figure. It is called as Under-modulation. Such a wave is called as an under-modulated wave.

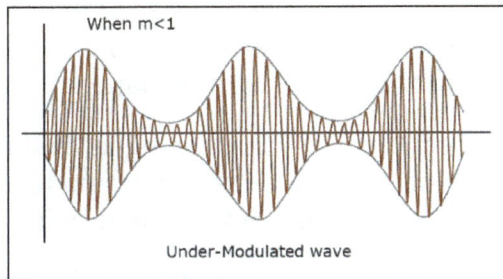

If the value of the modulation index is greater than 1, i.e., 1.5 or so, then the wave will be an over-modulated wave. It would look like the following figure.

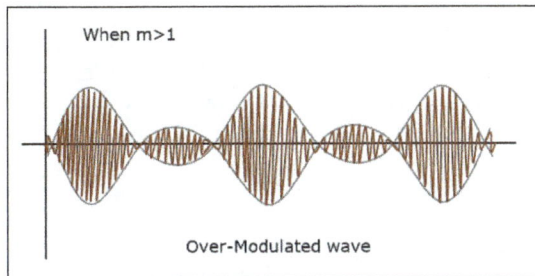

As the value of modulation index increases, the carrier experiences a 180° phase reversal, which causes additional sidebands and hence, the wave gets distorted. Such over modulated wave causes interference, which cannot be eliminated.

Bandwidth of Amplitude Modulation

The bandwidth is the difference between lowest and highest frequencies of the signal.

For amplitude modulated wave, the bandwidth is given by:

$$BW = f_{USB} - f_{LSB}$$

$$(f_c + f_m) - (f_c - f_m)$$

$$= 2f_m = 2W$$

Where, W is the message bandwidth.

Hence we got to know that the bandwidth required for the amplitude modulated wave is twice the frequency of the modulating signal.

Angle Modulation

Angle modulation is "the modulation in which phase or frequency of the carrier signal is varied with respect to message or transmitting signal". The angle modulation is classified as frequency modulation and phase modulation.

In frequency modulation, the frequency of carrier signal is altered in accordance with the the message or transmitting signal. The frequency modulation is also classified as narrow band width frequency modulation and wide band width frequency modulation.

In phase modulation, the phase angle of carrier signal is varied with respect to the message or transmitting signal. Consider frequency modulation, figure depicts the frequency modulation waveforms.

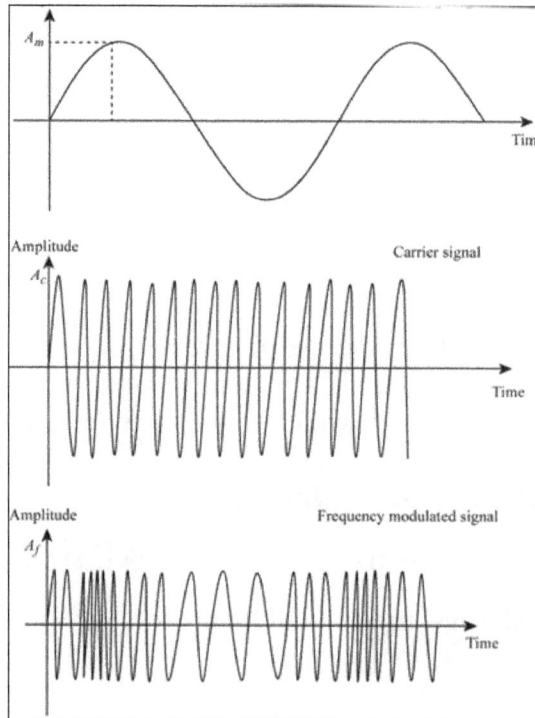

Consider a transmitting or message signal,

$$m(t) = A_m \cos(2\pi f_m t)$$

Here, A_m is amplitude of message signal and f_m is frequency of message signal.

Consider a carrier signal,

$$c(t) = A_c \cos(2\pi f_c t)$$

Here, A_c is amplitude of carrier signal and f_c is frequency of carrier signal.

Consider the frequency modulated signal can be given by,

$$f(t) = A_f \cos(2\pi f_c t + m \sin(2\pi f_m t))$$

Here, A_f is amplitude of frequency modulated signal and m is modulation index constant.

The narrow band width frequency modulation is used in ambulances, walkie – talkies, taxicabs and so on. The wide band frequency modulation is used in televisions, FM radios, and so on.

Frequency Modulation

Frequency modulation (FM) is a method of impressing data onto an alternating-current (AC) wave by varying the instantaneous frequency of the wave. This scheme can be used with analog or digital data.

In analog FM, the frequency of the AC signal wave, also called the *carrier*, varies in a continuous manner. Thus, there are infinitely many possible carrier frequencies. In *narrowband FM*, commonly used in two-way wireless communications, the instantaneous carrier frequency varies by up to 5 kilohertz (kHz, where 1 kHz = 1000 hertz or alternating cycles per second) above and below the frequency of the carrier with no modulation. In *wideband FM*, used in wireless broadcasting, the instantaneous frequency varies by up to several megahertz (MHz, where 1 MHz = 1,000,000 Hz). When the instantaneous input wave has positive polarity, the carrier frequency shifts in one direction; when the instantaneous input wave has negative polarity, the carrier frequency shifts in the opposite direction. At every instant in time, the extent of carrier-frequency shift (the *deviation*) is directly proportional to the extent to which the signal amplitude is positive or negative.

In digital FM, the carrier frequency shifts abruptly, rather than varying continuously. The number of possible carrier frequency states is usually a power of 2. If there are only two possible frequency states, the mode is called frequency-shift keying (FSK). In more complex modes, there can be four, eight, or more different frequency states. Each specific carrier frequency represents a specific digital input data state.

Frequency modulation is similar in practice to phase modulation (PM). When the instantaneous frequency of a carrier is varied, the instantaneous phase changes as well. The converse also holds: When the instantaneous phase is varied, the instantaneous frequency changes. But FM and PM are not exactly equivalent, especially in analog applications. When an FM receiver is used to demodulate a PM signal, or when an FM signal is intercepted by a receiver designed for PM, the audio is distorted. This is because the relationship between frequency and phase variations is not linear; that is, frequency and phase do not vary in direct proportion.

DSBSC Modulation

Every signal has to undergo modulation before sending it to the receiver. So, during the process of Amplitude Modulation, the modulated wave constitutes of the carrier wave and two sidebands. Sidebands contains information, Sideband is a band of frequencies, which contains power, with the higher and lower frequencies of the carrier frequency.

When a signal is transmitted which contains carrier wave with two sidebands then that is termed as Double Sideband Full Carrier system or simply DSBFC. It is plotted as shown in the following figure.

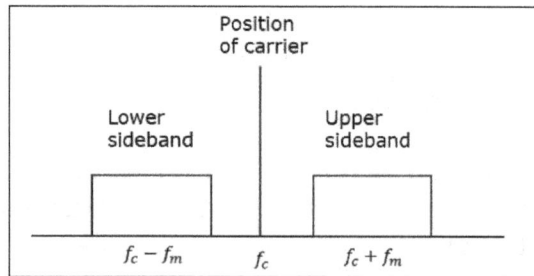

However, such a transmission is inefficient and ineffective. Because as we know already that carrier wave does not carry any information, then two-thirds of the power is being wasted in the carrier that exists no information.

If this carrier is going to be suppressed and this saved power is distributed between the two sidebands, then such a process is called as Double Sideband Suppressed Carrier system or simply DSBSC. It is plotted as shown in the following figure.

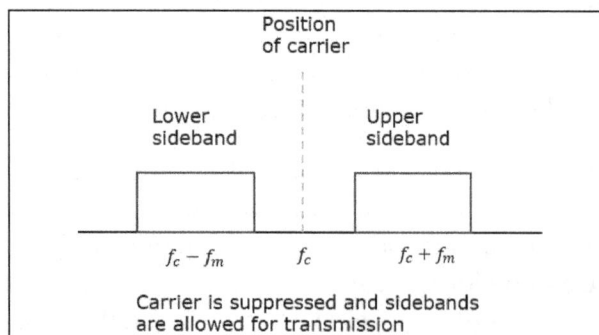

Mathematical Expressions

Let us consider the same mathematical expressions for modulating and carrier signals.

Modulating singnal,

$$m(t) = A_m \cos(2\pi f_m t)$$

and Carrier singnal,

$$c(t) = A_c \cos(2\pi f_c t)$$

Mathematically, we can represent the equation of DSBSC wave as the product of modulating and carrier singnals.

$$s(t) = m(t)c(t)$$
$$s(t) A_m A_c \cos(2\pi f_m t)\cos(2\pi f_c t)$$

Bandwidth of DSBSC Wave

We know the formula for bandwidth (BW) is,

$$BW = f_{max} - f_{min}$$

Consider the equation of DSBSC modulated wave.

$$s(t) = A_m A_c \cos(2\pi f_m t)\cos(2\pi f_c t)$$
$$s(t) = \frac{A_m A_c}{2}\cos\left[2\pi(f_c + f_m)t\right] + \frac{A_m A_c}{2}\cos\left[2\pi(f_c - f_m)t\right]$$

The DSBSC modulated wave has only two frequencies. So the maximum and minimum frequencies are $f_c + f_m$ and $f_c - f_m$ respectively.

i.e.,

$$f_{max} = f_c + f_m \text{ and } f_{min} = f_c - f_m$$

Substitute, f_{max} and f_{min} values in the bandwidth formula.

$$BW = f_c + f_m - (f_c - fm)$$
$$\Rightarrow BW = 2f_m$$

Thus, the bandwidth of DSBSC wave is same as that of AM wave and it is equal to twice the frequency of the modulating signal.

Power Calculations of DSBSC Wave

Consider the following equation of DSBSC modulated wave.

$$s(t) = \frac{A_m A_c}{2} \cos\left[2\pi(f_c + f_m)t\right] + \frac{A_m A_c}{2} \cos\left[2\pi(f_c - f_m)t\right]$$

Power of DSBSC wave is equal to the sum of powers of upper sideband and lower sideband frequency components.

$$P_t = P_{USB} + P_{LSB}$$

We know the standard formula for power of cos singnal is:

$$P = \frac{v_{rms}^2}{R} = \frac{(v_m \sqrt{2})^2}{R}$$

First, let us find the powers of upper sideband and lower sideband one by one upper sideband power:

$$P_{USB} = \frac{(A_m A_c / 2\sqrt{2})^2}{R} = \frac{A_m^2 A_c^2}{8R}$$

Similarly, we will get the lower sideband power same as that of upper sideband power.

$$P_{USB} = \frac{A_m^2 A_c^2}{8R}$$

Now, let us add these two sideband powers in order to get the power of DSBSC wave.

$$P_t = \frac{A_m^2 A_c^2}{8R} + \frac{A_m^2 A_c^2}{8R}$$

$$\Rightarrow P_t = \frac{A_m^2 A_c^2}{8R}$$

Therefore, the power required for transmitting DSBSC wave is equal to the power of both the sidebands.

DSBSC Modulators

The following two modulators generate DSBSC wave:

- Balanced modulator,
- Ring modulator.

Balanced Modulator

Balanced modulator consists of two identical AM modulators. These two modulators are arranged in a balanced configuration in order to suppress the carrier signal. Hence, it is called as balanced modulator.

Following is the block diagram of the balanced modulator.

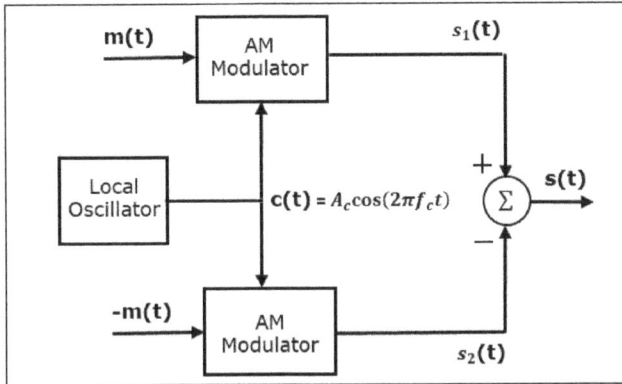

The same carrier signal $c(t) = A_c \cos(2\pi f_c t)$ is applied as one of the inputs to these two AM modulators. The modulating signal $m(t)$ is applied as another input to the upper AM modulator. Whereas, the modulating signal $m(t)$ ith opposite polarity, i.e., $-m(t)$ is applied as another input to the lower AM modulator.

Output of the upper AM modulator is,

$$s_1(t) = A_c[1 + k_a m(t)]\cos(2\pi f_c t)$$

Output of the lower AM modulator is,

$$s_2(t) = A_c[1 - k_a m(t)]\cos(2\pi f_c t)$$

We get the DSBSC wave $s(t)$ by subtracting $s_2(t)$ from $s_1(t)$. The summer block is used to perform this operation. $s_1(t)$ with positive sign and $s_2(t)$ with negative sign are applied as inputs to summer block. Thus, the summer block produces an output $s(t)$ which is the difference of $s_1(t)$ and $s_2(t)$.

$$s(t) = A_c[1 + k_a m(t)]\cos(2\pi f_c t) - A_c[1 - k_a m(t)]\cos(2\pi f_c t)$$
$$s(t) = A_c \cos(2\pi f_c t) + A_c k_a m(t)\cos(2\pi f_c t) - A_c \cos(2\pi f_c t) + A_c k_a m(t)\cos(2\pi f_c t)$$

$$s(t) = 2A_c k_a m(t)\cos(2\pi f_c t)$$

We know the standard equation of DSBSC wave is,

$$s(t) = A_c m(t)\cos(2\pi f_c t)$$

By comparing the output of summer block with the standard equation of DSBSC wave, we will get the scaling factor as $2k_a$.

Ring Modulator

Following is the block diagram of the Ring modulator.

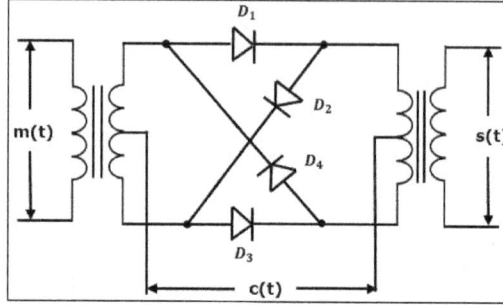

In the diagram, the four diodes D_1, D_2, D_3 and D_4 are connected in the ring structure. Hence, this modulator is called as the ring modulator. Two center tapped transformers are used in this diagram. The message signal $m(t)$ is applied to the input transformer. Whereas, the carrier signals $c(t)$ is applied between the two center tapped transformers.

For positive half cycle of the carrier signal, the diodes D_1 and D_3 are switched ON and the other two diodes D_2 and D_4 are switched OFF. In this case, the message signal is multiplied by +1.

For negative half cycle of the carrier signal, the diodes D_2 and D_4 are switched ON and the other two diodes D_1 and D_3 are switched OFF. In this case, the message signal is multiplied by -1. This results in $180°$ phase shift in the resulting DSBSC wave.

From the above analysis, we can say that the four diodes D_1, D_2, D_3 and D_4 are controlled by the carrier signal. If the carrier is a square wave, then the Fourier series representation of $c(t)$ is represented as,

$$c(t) = \frac{4}{\pi} \sum_{n=1}^{\infty} \frac{(-1)^{n-1}}{2n-1} \cos[2\pi f_c t(2n=-1)]$$

We will get DSBSC wave $s(t)$, which is just the product of the carrier signal $c(t)$ and the message signal $m(t)$ i.e.,

$$s(t) = \frac{4}{\pi} \sum_{n=1}^{\infty} \frac{(-1)^{n-1}}{2n-1} \cos[2\pi f_c t)(2n-1)]m(t)$$

The above equation represents DSBSC wave, which is obtained at the output transformer of the ring modulator.

DSBSC modulators are also called as product modulators as they produce the output, which is the product of two input signals.

DSBSC Demodulators

DSBSC is an amplitude modulation basically without a carrier. It is the process of detecting the signal which was modulated by using double sideband suppressed-carrier amplitude modulation. The following are the two demodulators (detectors) used for demodulating DSBSC wave:

- Coherent Detector,
- Costas Loop.

Coherent Detector

Here, we use the same carrier signal (which is used for generating DSBSC signal) is applied to detect the message signal. Hence, this process of detection is called as coherent or synchronous detection. Following is the block diagram of the coherent detector.

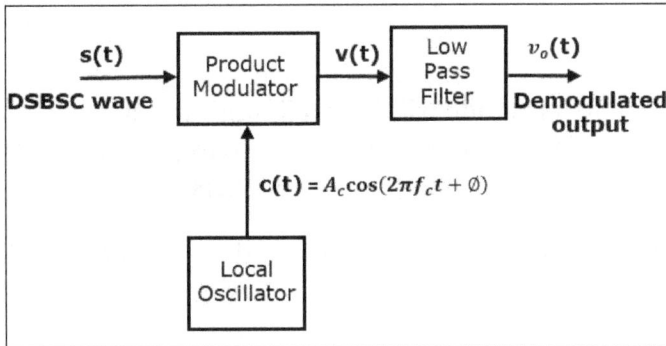

In this process, the message signal can be extracted from DSBSC wave by multiplying it with a carrier, with the same frequency and the phase of the carrier that is used in DSBSC modulation. The output of the resulting signal is then passed through a Low Pass Filter. Output of this filter is the confirmed as the desired message signal.

Let the DSBSC wave be,

$$s(t) = A_c \cos(2\pi f_c t)m(t)$$

The output of the local oscillator is,

$$c(t) = A_c \cos(2\pi f_c t + \phi)$$

Where, ϕ is the phase difference between the local oscillator signal and the carrier signal, which is used for DSBSC modulation.

From the figure, we can write the output of product modulator as,

$$v(t) = s(t)c(t)$$

Substitute, $s(t)$ and $c(t)$ values in the above equation.

$$v(t) = A_c \cos(2\pi f_c t)m(t)A_c \cos(f_c t + \phi)$$
$$= A_c^2 \cos(2\pi f_c t)\cos(2\pi f_c t + \phi)m(t)$$
$$= \frac{A_c^2}{2}[\cos(4\pi f_c t + \phi) + \cos\phi]m(t)$$
$$v(t) = \frac{A_c^2}{2}\cos\phi m(t) + \frac{A_c^2}{2}\cos(4\pi f_c t + \phi)m(t)$$

In the above equation, the first term is considered as the scaled version of the message signal. It can be extracted by passing the above signal through a low pass filter.

Therefore, the output of low pass filter is,

$$v_0 t = \frac{A_c^2}{2}\cos\phi m(t)$$

The demodulated signal amplitude will be maximum, when $\phi = 0^0$ that's why both the local oscillator signal and the carrier signal must be in phase, so that there should not appear any phase difference between these two signals.

The demodulated signal amplitude will be zero, when $\phi = \pm 90°$ this effect is called as quadrature null effect.

Costas Loop

Costas loop consists of both the carrier signal (used for DSBSC modulation) and the locally generated signal which are in phase. Following is the block diagram of Costas loop.

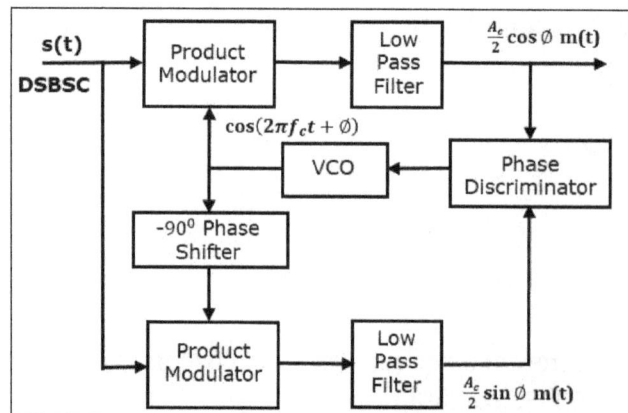

By seeing the diagram, you can find the Costas loop which consists of two product modulators with common input s(t)s(t), which is DSBSC wave to both the modulators. The second input for both product modulators is taken from Voltage Controlled Oscillator (VCO) with $-90°$ phase shift to one of the product modulator as shown in figure.

We know that the equation of DSBSC wave is,

$$s(t) = A_c \cos(2\pi f_c t)m(t)$$

Let the output of VCO be,

$$c_1(t) = \cos(2\pi f_c t + \phi)$$

This output of VCO is applied as the carrier input of the upper product modulator.

Hence, the output of the upper product modulator is,

$$v_1(t) = s(t)c_1(t)$$

Substitute, $s(t)$ and $c_1(t)$ values in the above equation.

$$v_1(t) = A_c \cos(2\pi f_c t)m(t)\cos(2\pi f_c t + \phi)$$

After simplifying, we will get $v_1(t)$ as,

$$v_1(t) = \frac{A_c}{2}\cos\phi\, m(t) + \frac{A_c}{2}\cos(4\pi f_c t + \phi)m(t)$$

This signal is applied as an input of the upper low pass filter. The output of the low pass filter is,

$$v_{01}(t) = \frac{A_c}{2}\cos\phi\, m(t)$$

Therefore, the output of this low pass filter is the scaled version of the modulating signal.

The output of $-90°$ phase shifter is,

$$c_2(t) = \cos(2\pi f_c t + \phi - 90°) = \sin(2\pi f_c t + \phi)$$

This signal is applied as the carrier input of the lower product modulator.

The output of the lower product modulator is,

$$v_2(t) = s(t)c_2(t)$$

Substitute, $s(t)$ and $c_2(t)$ values in the above equation.

$$v_2(t) = A_c \cos(2\pi f_c t)m(t)\sin(2\pi f_c t + \phi)$$

After simplifying, we will get $v_2(t)$ as,

$$v_2(t) = \frac{A_c}{2}\sin\phi m(t) + \frac{A_c}{2}\sin(4\pi f_c t + \phi)m(t)$$

This signal is applied as an input of the lower low pass filter. The output of this low pass filter is,

$$v_{02}(t) = \frac{A_c}{2}\sin\phi m(t)$$

The output of this Low pass filter has −900−900 phase difference with the output of the upper low pass filter.

The outputs of these two low pass filters that we get are applied as inputs of the phase discriminator. Depending upon the phase difference between these two signals, the phase discriminator produces a DC control signal.

This signal is further applied as an input of VCO to correct the phase error in VCO output. Therefore, the carrier signal (used for DSBSC modulation) and the locally generated signal (VCO output) are considered to be in phase. It is mandatory that both signals to be in phase.

SSBSC Modulation

The process of suppressing one of the sidebands along with the carrier and transmitting a single sideband is called as Single Sideband Suppressed Carrier system or simply SSBSC. It is plotted as shown in the following figure.

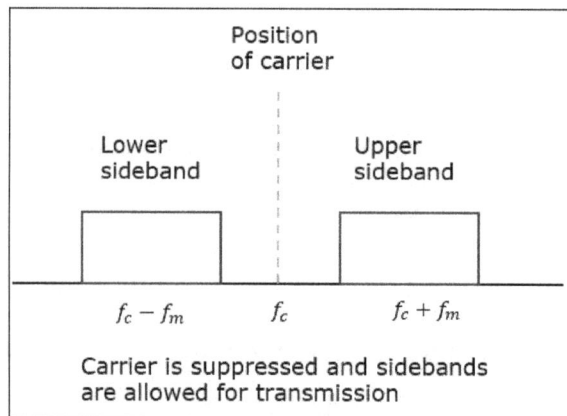

In the above figure, the carrier and the lower sideband are suppressed. Hence, the upper sideband is used for transmission. Similarly, we can suppress the carrier and the upper sideband while transmitting the lower sideband.

This SSBSC system, which transmits a single side band has high power, as the power allotted for both the carrier and the other sideband is utilized in transmitting this Single Sideband.

Mathematical Expressions

Let us consider the same mathematical expressions for the modulating and the carrier signals.

Modulating signal,

$$m(t) = A_m \cos(2\pi f_m t)$$

and Carrier signal,

$$c(t) = A_c \cos(2\pi f_c t)$$

Mathematically, we can represent the equation of SSBSC wave as,

$$s(t) = \frac{A_m A_c}{2} \cos\left[2\pi(f_c + f_m)t\right]$$

for the upper sideband or for the lower sideband,

$$s(t) = \frac{A_m A_c}{2} \cos\left[2\pi(f_c - f_m)t\right]$$

Bandwidth of SSBSC Wave

We know that the DSBSC modulated wave contains two sidebands and its bandwidth is $2f_m$ Since the SSBSC modulated wave contains only one sideband, its bandwidth is half of the bandwidth of DSBSC modulated wave.

i.e., Bandwidth of SSBSC modulated wave = $\dfrac{2f_m}{2} = f_m$

Therefore, the bandwidth of SSBSC modulated wave is f_m and it is equal to the frequency of the modulating signal.

Power Calculations of SSBSC Wave

Consider the following equation of SSBSC modulated wave.

$$s(t) = \frac{A_m A_c}{2} \cos\left[2\pi(f_c + f_m)t\right]$$

for the upper sideband,

or

$$s(t) = \frac{A_m A_c}{2} \cos\left[2\pi(f_c - f_m)t\right]$$

Power of SSBSC wave is equal to the power of any one sideband frequency components.

$$P_t = P_{USB} = P_{LSB}$$

We know that the standard formula for power of cos signal is,

$$P = \frac{v_{rms}^2}{R} = \frac{\left(v_m / \sqrt{2}\right)^2}{R}$$

In this case, the power of the upper sideband is,

$$P_{USB} \frac{\left(A_m A_c / 2\sqrt{2}\right)^2}{R} = \frac{A_m^2 A_c^2}{8R}$$

Similarly, we will get the lower sideband power same as that of the upper side band power.

$$P_{LSB} = \frac{A_m^2 A_c^2}{8R}$$

Therefore, the power of SSBSC wave is,

$$P_t = P_{USB} = P_{LSB} = \frac{A_m^2 A_c^2}{8R}$$

Advantages

- Bandwidth or spectrum space occupied is lesser than AM and DSBSC waves.
- Transmission of more number of signals is allowed.
- Power is saved.
- High power signal can be transmitted.
- Less amount of noise is present.
- Signal fading is less likely to occur.

Disadvantages

- The generation and detection of SSBSC wave is a complex process.

- The quality of the signal gets affected unless the SSB transmitter and receiver have excellent frequency stability.

Applications

- For power saving requirements and low bandwidth requirements.

- In land, air, and maritime mobile communications.

- In point-to-point communications.

- In radio communications.

- In television, telemetry, and radar communications.

- In military communications, such as amateur radio, etc.

SSBSC Modulators

Two methods are used to generate SSBSC wave, they are as follows:

- Frequency discrimination method.

- Phase discrimination method.

Frequency Discrimination Method

The following figure shows the block diagram of SSBSC modulator using frequency discrimination method.

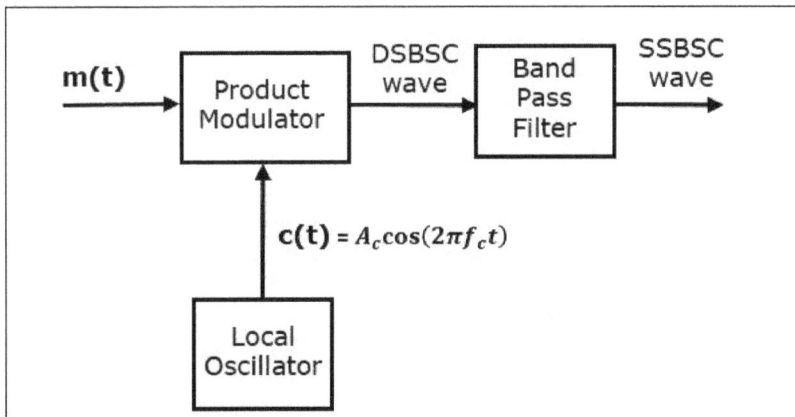

In this method, you can see that DSBSC wave is generated with the help of the product modulator. Then, DSBSC wave is applied as an input of band pass filter. With passing the input, the band pass filter produces an output, results in SSBSC wave.

This band pass filter is set to frequency range as the spectrum of the desired SSBSC wave which can be tuned either to upper sideband or lower sideband frequencies so that respective SSBSC wave is generated with upper sideband or lower sideband.

Phase Discrimination Method

The following figure shows the block diagram of SSBSC modulator using phase discrimination method.

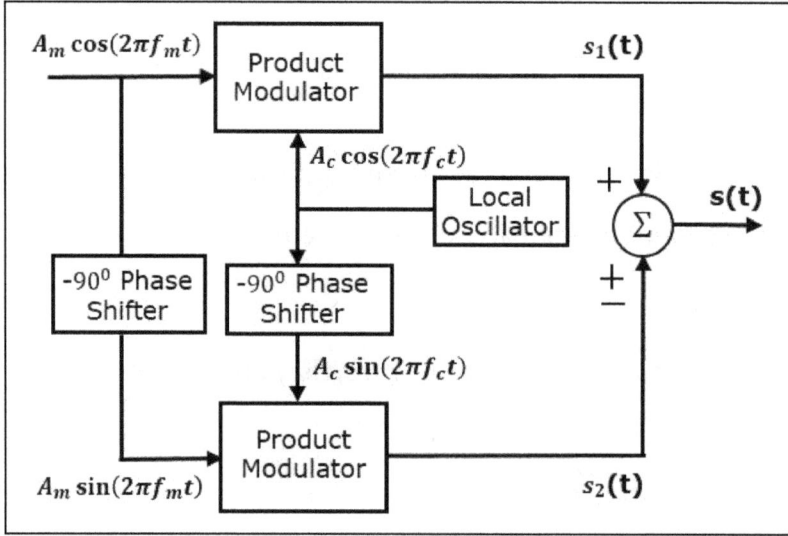

The above block diagram consists of two product modulators that is upper product modulator and lower product modulator, two −90° phase shifters, one is local oscillator and one is summer block. The first product modulator produces an output, which is the product of two inputs. The −90° hase shifter produces an output, having a phase lag of −90° with respect to the input.

In the two phase shifters the first one called local oscillator is used to produce the carrier signal. Summer block produces an output, which may be either the sum of two inputs or the difference between two inputs depending on the polarity of inputs.

The modulating signal $A_m \cos(2\pi f_m t)$ and the carrier signal $A_c \cos(2\pi f_c t)$ are directly applied as inputs to the upper product modulator. So, the upper product modulator generates an output, which is the resultant of these two inputs.

The output of upper product modulator is,

$$s_1(t) = A_m A_c \cos(2\pi f_m t)\cos(2\pi f_c t)$$
$$s_1(t) = \frac{A_m A_c}{2}\{\cos[2\pi(f_c + f_m)t] + \cos[2\pi(f_c - f_m)t]\}$$

The modulating signal $A_m \cos(2\pi f_m t)$ and the carrier signal $A_c \cos(2\pi f_c t)$ are phase shifted by $-90°$ before applying as inputs to the lower product modulator. So the lower product modulator produces an output, which is the resultant of these two inputs.

The output of lower product modulator is,

$$s_2(t) = A_m A_c \cos(2\pi f_m t - 90°) \cos(2\pi f_c t - 90°)$$
$$s_2(t) A_m A_c \sin(2\pi f_m t) \sin(2\pi f_c t)$$
$$s_2(t) = \frac{A_m A_c}{2} \{\cos[2\pi(f_c - f_m)t] - \cos[2\pi(f_c + f_m)t]\}$$

Add $s_1(t)$ and $s_2(t)$ in order to get the SSBSC modulated wave $s(t)$ having a lower sideband.

$$s(t) = \frac{A_m A_c}{2} \{\cos[2\pi(f_c + f_m)t] + \cos[2\pi(f_c - f_m)t]\} +$$

$$\frac{A_m A_c}{2} \{\cos[2\pi(f_c - f_m)t] - \cos[2\pi(f_c + f_m)t]\}$$
$$s(t) = A_m A_c \cos[2\pi(f_c - f_m)t]$$

Subtract $s_2(t)$ from $s_1(t)$ in order to get the SSBSC modulated wave $s(t)$ having a upper sideband.

$$s(t) = \frac{A_m A_c}{2} \{\cos[2\pi(f_c + f_m)t] + \cos[2\pi(f_c - f_m)t]\} -$$

$$\frac{A_m A_c}{2} \{\cos[2\pi(f_c - f_m)t] - \cos[2\pi(f_c + f_m)t]\}$$
$$s(t) = A_m A_c \cos[2\pi(f_c - f_m)t]$$

Hence, based on the polarities of input that we have chosen as inputs at summer block, we will get the desired SSBSC wave having an upper sideband or a lower sideband.

SSBSC Demodulator

The process of extracting an original message signal from SSBSC wave is known as detection or demodulation of SSBSC. Coherent detector is used for demodulating SSBSC wave.

Coherent Detector

Here, the same carrier signal (which is used for generating SSBSC wave) is used to detect the message signal. Hence, this process of detection is called as coherent or synchronous detection. Following is the block diagram of coherent detector.

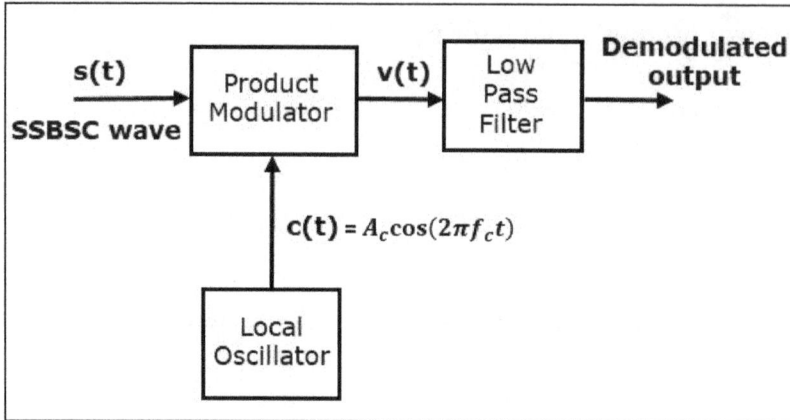

In this process, the message signal can be extracted from SSBSC wave by multiplying it with a carrier, having the same frequency and the phase of the carrier used in SSBSC modulation. The resulting signal is then passed through a Low Pass Filter. The output of this filter is the desired message signal.

Consider the following SSBSC wave having a lower sideband.

$$s(t) = \frac{A_m A_c}{2} \cos[2\pi(f_c - f_m)t]$$

The output of the local oscillator is,

$$c(t) = A_c \cos(2\pi f_c t)$$

From the figure, we can write the output of product modulator as,

$$v(t) = s(t)c(t)$$

Substitute $s(t)$ and $c(t)$ values in the above equation.

$$v(t) = \frac{A_m A_c}{2} \cos[2\pi f_c - f_m)t]A_c \cos(2\pi f_c t)$$

$$= \frac{A_m A_c}{2} \cos[2\pi(f_c - f_m)t]\cos(2\pi f_c t)$$

$$= \frac{A_m A_c^2}{4}\{\cos[2\pi(2f_c - f_m)] + \cos(2\pi f_m)t\}$$

$$v(t) = \frac{A_m A_c^2}{4}\cos(2\pi f_m t) + \frac{A_m A_c^2}{4}\cos[2\pi(2f_c - f_m)t]$$

In the equation, the first term is the scaled version of the message signal. It can be extracted by passing the above signal through a low pass filter.

Therefore, the output of low pass filter is,

$$v_0(t) = \frac{A_m A_c}{4} \cos(2\pi f_m t)$$

Here, the scaling factor is $\frac{A_c^2}{4}$.

We can use the same block diagram for demodulating SSBSC wave having an upper sideband. Consider the following SSBSC wave having an upper sideband.

$$s(t) = \frac{A_m A_c}{2} \cos[2\pi(f_c + f_m)t]$$

The output of the local oscillator is,

$$c(t) = A_c \cos(2\pi f_c t)$$

We can write the output of the product modulator as,

$$v(t) = s(t) c(t)$$

Substitute $s(t)$ and $c(t)$ values in the above equation.

$$v(t) = \frac{A_m A_c}{2} \cos[2\pi(f_c + f_m)t] A_c \cos(2\pi f_c t)$$

$$= \frac{A_m A_c^2}{2} \cos[2\pi(f_c + f_m)t] \cos(2\pi f_c t)$$

$$= \frac{A_m A_c^2}{4} \{\cos[2\pi(2f_c + f_m)t] + \cos(2\pi f_m t)\}$$

$$v(t) = \frac{A_m A_c^2}{4} \cos(2\pi f_m t) + \frac{A_m A_c^2}{4} \cos[2\pi(2f_c + f_m)t]$$

In the above equation, the first term is the scaled version of the message signal. It can be extracted by passing the above signal through a low pass filter.

Therefore, the output of the low pass filter is,

$$v_0(t0 = \frac{A_m A_c^2}{4} \cos(2\pi f_m t)$$

Here too the scaling factor is $\frac{A_c^2}{4}$.

Therefore, we get the same demodulated output in both the cases by using coherent detector.

VSBSC Modulation

In order to avoid this loss of information, a technique is chosen, which is known as Vestigial Side Band Suppressed Carrier (VSBSC) technique. The word "vestige" means "a certain part" from which, the name is derived.

VSBSC Modulation is an amplitude modulation technique, where one of the side band is removed to form a vestige modulation hence, the name vestige modulation. In other words a part of the signal is modulated along with one sideband. The frequency spectrum of VSBSC wave is shown in the following figure.

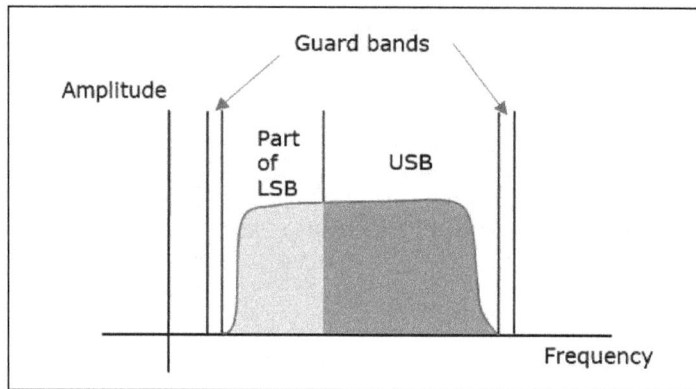

With the two sidebands, the upper sideband and also a part of the lower sideband is also being transmitted in this technique. It is quite opposite with the lower sideband, it transmits the lower sideband along with a part of the upper sideband. To avoid the interference between the bands a small width of guard band is laid on either side of VSB the. VSB modulation is mostly used in television transmissions.

Bandwidth of VSBSC Modulation

We know that the bandwidth of SSBSC modulated wave is fmfm. Since the VSBSC modulated wave contains the frequency components of one side band along with the vestige of other sideband, the bandwidth of it will be the sum of the bandwidth of SSBSC modulated wave and vestige frequency fv. i.e., bandwidth of VSBSC modulated Wave = fm + fv.

Advantages

Following are the advantages of VSBSC modulation:

- Highly efficient.

- Bandwidth is reduced when compared to AM and DSBSC waves.

- Filter design is easy, as that doesn't require high accuracy.

- The transmission of low frequency components is possible, without any difficulty.

- Possesses good phase characteristics.

Disadvantages

Following are the disadvantages of VSBSC modulation:

- More Bandwidth when compared to SSBSC wave.

- Demodulation is complex.

Applications

The most prominent and standard application of VSBSC is for the transmission of television signals. Also, this is the most convenient and efficient technique in the case of bandwidth usage.

Now, let us discuss about the modulator which generates VSBSC wave and the demodulator which demodulates VSBSC wave one by one.

Generation of VSBSC

Generation of VSBSC wave is similar to the generation of SSBSC wave. The VSBSC modulator is shown in the figure.

In this method, first we will generate DSBSC wave with the help of the product modulator. Then, apply this DSBSC wave as an input of sideband shaping filter. This filter produces an output, which is VSBSC wave.

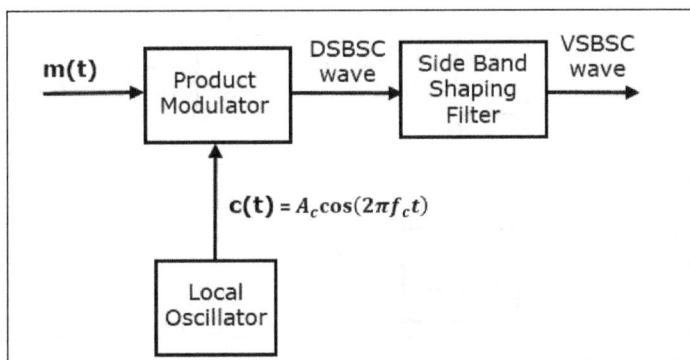

The modulating signal m(t) and carrier signal A cos(2πfct) and cos(2πfct) are applied as inputs to the product modulator. Hence, the product modulator produces an output, which is the product of these two inputs.

Therefore, the output of the product modulator is,

$$p(t) = A_c \cos(2\pi f_c t) m(t)$$

Apply Fourier transform on both sides,

$$P(f) = \frac{A_c}{2}[M(f - f_c) + M(f + f_c)]$$

The above equation represents the equation of DSBSC frequency spectrum.

Let the transfer function of the sideband shaping filter be $H(f)$. this filter has the input $p(t)$ and the output is VSBSC modulated wave $s(t)$. The Fourier transforms of $p(t)$ and $s(t)$ are $P(t)$ and $S(t)$. respectively.

Mathematically, we can write $S(f)$ as,

$$s(t) = P(f)H(f)$$

Substitute $P(f)$ value in the above equation.

$$S(f) = \frac{A_c}{2}[M(f - f_c) + M(f + f_c)]H(f)$$

Demodulation of VSBSC

Demodulation of VSBSC wave is similar to the demodulation of SSBSC wave. Here, the same carrier signal (which is used for producing VSBSC wave) is used to detect the message signal. Hence, this process of detection is called as coherent or synchronous detection. The VSBSC demodulator is shown in the figure.

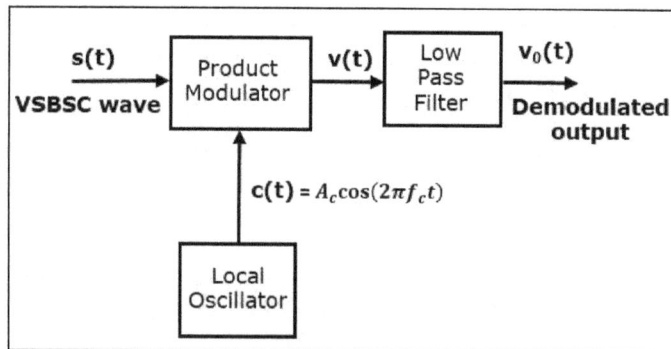

In this process, you can get the message signal from VSBSC wave by extraction when it is multiplied it with a carrier, which is having the same frequency and the carrier phase used in VSBSC modulation. The resulting signal is then passed through a Low Pass Filter. The output of this filter is the desired message signal.

Let the VSBSC wave be $s(t)$ and the carrier signal is $A_c \cos(2\pi f_c t)$.

From the figure, we can write the output of the product modulator as:

$$v(t) = A_c \cos(2\pi f_c t) s(t)$$

Apply Fourier transform on both sides,

$$V(f) = \frac{A_c}{2}[S(f - f_c) + S(f + f_c)$$

We know that, $S(f) = \frac{A_c}{2}[M(f - f_c) + M(f + f_c)]H(f)$

From the above equation, let us find $S(f - f_c)$ and $S(f + f_c)$.

$$S(f - f_c) = \frac{A_c}{2}[M(f - f_c - f_c) + M(f - f_c + f_c)]H(f - f_c)$$

$$\Rightarrow S(f - f_c) + \frac{A_c}{2}[M(f - 2f_c) + M(f)]H(f - f_c)$$

$$S(f + f_c) = \frac{A_c}{2}[M(f + f_c - f_c) + M(f + f_c + f_c)]H(f + f_c)$$

$$\Rightarrow S(f + f_c) = \frac{A_c}{2}[M(f) + M(f + 2f_c)]H(f + f_c)$$

Substitute $S(f - f_c)$ and $S(f + f_c)$ values $V(f)$.

$$V(f) = \frac{A_c}{2}[\frac{A_c}{2}[M(f - 2f_c) + M(f)]H(f - f_c) + \frac{A_c}{2}[M(f) + M(+2f_c)]H(f + f_c)]$$

$$\Rightarrow V(f) = \frac{A_c^2}{4}M(f)[H(f - f_c) + H(f + f_c)] + \frac{A_c^2}{4}[M(f - 2f_c)H(f - f_c) +$$

$$M(f + 2f_c)H(f + f_c)]$$

In the above equation, the first term represents the scaled version of the desired message signal frequency spectrum. It can be extracted by passing the above signal through a low pass filter.

$$V_0(f) = \frac{A_c^2}{4}M(f)[H(f - f_c) + H(f + f_c)]$$

References

- Analog-communication: ukessays.com, Retrieved 11 June, 2019
- Analog-communication-modulation: tutorialspoint.com, Retrieved 13 January, 2019

- Principles-of-communication-amplitude-modulation: tutorialspoint.com, Retrieved 3 May, 2019
- Angle-modulation-4: chegg.com, Retrieved 23 August, 2019
- Frequency-modulation: techtarget.com, Retrieved 7 July, 2019
- Analog-communication-dsbsc-modulation-25603: wisdomjobs.com, Retrieved 18 March, 2019
- Analog-communication-dsbsc-modulators: tutorialspoint.com, Retrieved 3 February, 2019
- Dsbsc-demodulators-25605: wisdomjobs.com, Retrieved 20 April, 2019
- Analog-communication-ssbsc-modulation: tutorialspoint.com, Retrieved 25 July, 2019
- Analog-communication-ssbsc-modulators-25606: wisdomjobs.com, Retrieved 8 June, 2019
- Analog-communication-ssbsc-demodulator: tutorialspoint.com, Retrieved 13 March, 2019
- Analog-communication-vsbsc-modulation-25608: wisdomjobs.com, Retrieved 30 August, 2019
- Analog-communication-multiplexing: tutorialspoint.com, Retrieved 27 April, 2019

3

Digital Communications and Transmission

The mode of communication where information is encoded in a digital format and then transferred electronically is known as digital communication. This chapter discusses in detail the components and techniques related to digital communications and transmission such as sampling techniques, digital modulation techniques and M-ary encoding.

Digital Communications

Necessity of Digitization

The conventional methods of communication used analog signals for long distance communications, which suffer from many losses such as distortion, interference, and other losses including security breach.

In order to overcome these problems, the signals are digitized using different techniques. The digitized signals allow the communication to be more clear and accurate without losses.

The following figure indicates the difference between analog and digital signals. The digital signals consist of 1s and 0s which indicate High and Low values respectively.

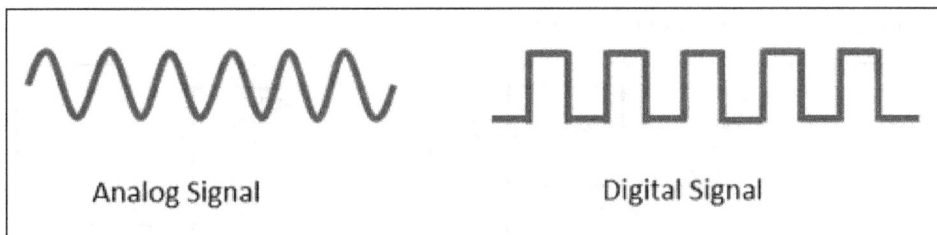

Analog Signal   Digital Signal

Advantages

As the signals are digitized, there are many advantages of digital communication over analog communication, such as:

- The effect of distortion, noise, and interference is much less in digital signals as they are less affected.

- Digital circuits are more reliable.

- Digital circuits are easy to design and cheaper than analog circuits.

- The hardware implementation in digital circuits is more flexible than analog.

- The occurrence of cross-talk is very rare in digital communication.

- The signal is un-altered as the pulse needs a high disturbance to alter its properties, which is very difficult.

- Signal processing functions such as encryption and compression are employed in digital circuits to maintain the secrecy of the information.

- The probability of error occurrence is reduced by employing error detecting and error correcting codes.

- Spread spectrum technique is used to avoid signal jamming.

- Combining digital signals using Time Division Multiplexing (TDM) is easier than combining analog signals using Frequency Division Multiplexing (FDM).

- The configuring process of digital signals is easier than analog signals.

- Digital signals can be saved and retrieved more conveniently than analog signals.

- Many of the digital circuits have almost common encoding techniques and hence similar devices can be used for a number of purposes.

- The capacity of the channel is effectively utilized by digital signals.

Elements of Digital Communication

The elements which form a digital communication system is represented by the following block diagram for the ease of understanding.

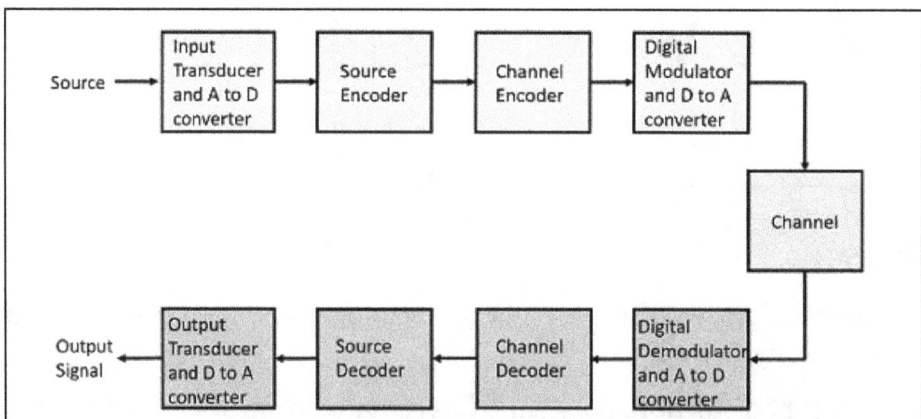

Basic elements of digital communication system.

Following are the sections of the digital communication system.

Input Transducer

This is a transducer which takes a physical input and converts it to an electrical signal (Example: microphone). This block also consists of an analog to digital converter where a digital signal is needed for further processes.

A digital signal is generally represented by a binary sequence.

Source Encoder

The source encoder compresses the data into minimum number of bits. This process helps in effective utilization of the bandwidth. It removes the redundant bits (unnecessary excess bits, i.e., zeroes).

Channel Encoder

The channel encoder does the coding for error correction. During the transmission of the signal, due to the noise in the channel, the signal may get altered and hence to avoid this, the channel encoder adds some redundant bits to the transmitted data. These are the error correcting bits.

Digital Modulator

The signal to be transmitted is modulated here by a carrier. The signal is also converted to analog from the digital sequence, in order to make it travel through the channel or medium.

Channel

The channel or a medium, allows the analog signal to transmit from the transmitter end to the receiver end.

Digital Demodulator

This is the first step at the receiver end. The received signal is demodulated as well as converted again from analog to digital.

Channel Decoder

The channel decoder, after detecting the sequence, does some error corrections. The distortions which might occur during the transmission are corrected by adding some redundant bits. This addition of bits helps in the complete recovery of the original signal.

Source Decoder

The resultant signal is once again digitized by sampling and quantizing so that the pure digital output is obtained without the loss of information. The source decoder recreates the source output.

Output Transducer

This is the last block which converts the signal into the original physical form, which was at the input of the transmitter. It converts the electrical signal into physical output (Example: loud speaker).

Output Signal

This is the output which is produced after the whole process. Example – The sound signal received.

This unit has dealt with the introduction, the digitization of signals, the advantages and the elements of digital communications.

Pulse Code Modulation

Modulation is the process of varying one or more parameters of a carrier signal in accordance with the instantaneous values of the message signal. The message signal is the signal which is being transmitted for communication and the carrier signal is a high frequency signal which has no data, but is used for long distance transmission.

There are many modulation techniques, which are classified according to the type of modulation employed. Of them all, the digital modulation technique used is Pulse Code Modulation (PCM).

A signal is pulse code modulated to convert its analog information into a binary sequence, i.e., 1s and 0s. The output of a PCM will resemble a binary sequence. The following figure shows an example of PCM output with respect to instantaneous values of a given sine wave.

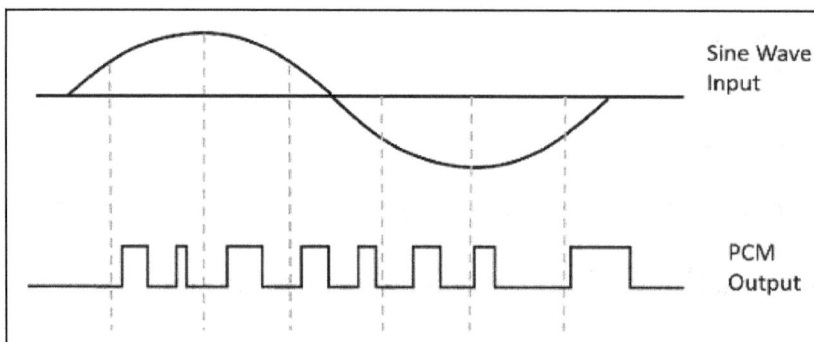

Instead of a pulse train, PCM produces a series of numbers or digits, and hence this process is called as digital. Each one of these digits, though in binary code, represents the approximate amplitude of the signal sample at that instant.

In Pulse Code Modulation, the message signal is represented by a sequence of coded pulses. This message signal is achieved by representing the signal in discrete form in both time and amplitude.

Basic Elements of PCM

The transmitter section of a Pulse Code Modulator circuit consists of Sampling, Quantizing and Encoding, which are performed in the analog-to-digital converter section. The low pass filter prior to sampling prevents aliasing of the message signal.

The basic operations in the receiver section are regeneration of impaired signals, decoding, and reconstruction of the quantized pulse train. Following is the block diagram of PCM which represents the basic elements of both the transmitter and the receiver sections.

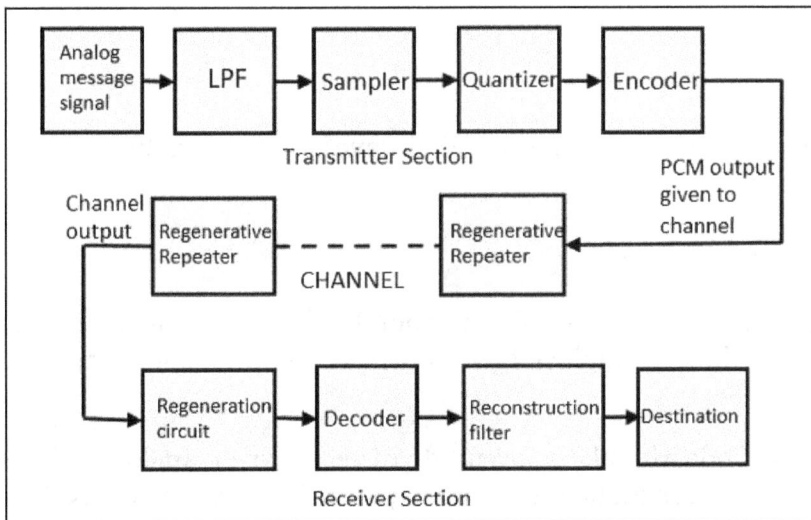

Low Pass Filter

This filter eliminates the high frequency components present in the input analog signal which is greater than the highest frequency of the message signal, to avoid aliasing of the message signal.

Sampler

This is the technique which helps to collect the sample data at instantaneous values of message signal, so as to reconstruct the original signal. The sampling rate must be greater than twice the highest frequency component W of the message signal, in accordance with the sampling theorem.

Quantizer

Quantizing is a process of reducing the excessive bits and confining the data. The sampled output when given to Quantizer reduces the redundant bits and compresses the value.

Encoder

The digitization of analog signal is done by the encoder. It designates each quantized level by a binary code. The sampling done here is the sample-and-hold process. These three sections (LPF, Sampler, and Quantizer) will act as an analog to digital converter. Encoding minimizes the bandwidth used.

Regenerative Repeater

This section increases the signal strength. The output of the channel also has one regenerative repeater circuit, to compensate the signal loss and reconstruct the signal, and also to increase its strength.

Decoder

The decoder circuit decodes the pulse coded waveform to reproduce the original signal. This circuit acts as the demodulator.

Reconstruction Filter

After the digital-to-analog conversion is done by the regenerative circuit and the decoder, a low-pass filter is employed, called as the reconstruction filter to get back the original signal.

Hence, the Pulse Code Modulator circuit digitizes the given analog signal, codes it and samples it, and then transmits it in an analog form. This whole process is repeated in a reverse pattern to obtain the original signal.

Digital Communication

For the samples that are highly correlated, when encoded by PCM technique, leave redundant information behind. To process this redundant information and to have a better output, it is a wise decision to take a predicted sampled value, assumed from its previous output and summarize them with the quantized values. Such a process is called as Differential PCM (DPCM) technique.

DPCM Transmitter

The DPCM Transmitter consists of Quantizer and Predictor with two summer circuits.

Following is the block diagram of DPCM transmitter.

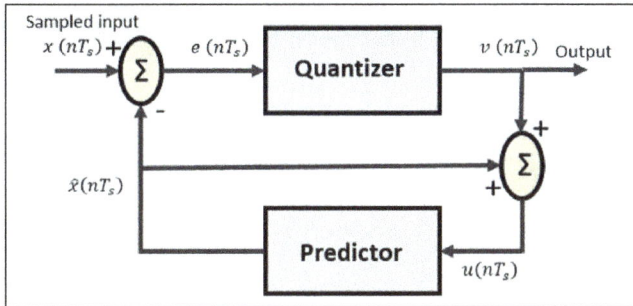

The signals at each point are named as:

- $x(nTs)$ is the sampled input;

- $\hat{x}(nT_s)$ is the predicted sample;

- $e(nT_s)$ is the difference of sampled input and predicted output, often called as prediction error;

- $v(nT_s)$ is the quantized output;

- $u(nT_s)$ is the predictor input which is actually the summer output of the predictor output and the quantizer output.

The predictor produces the assumed samples from the previous outputs of the transmitter circuit. The input to this predictor is the quantized versions of the input signal $x(nT_s)$.

Quantizer Output is represented as:

$$v(nT_s) = Q\big[e(nT_s)\big]$$

$$= e(nT_s) + q(nT_s)$$

Where, q (nT$_s$) is the quantization error.

Predictor input is the sum of quantizer output and predictor output,

$$u(nT_s) = \hat{x}(nT_s) + v(nT_s)$$

$$u(nT_s) = \hat{x}(nT_s) + e(nT_s) + q(nT_s)$$

$$u(nT_s) = x(nT_s) + q(nT_s)$$

The same predictor circuit is used in the decoder to reconstruct the original input.

DPCM Receiver

The block diagram of DPCM Receiver consists of a decoder, a predictor, and a summer circuit.

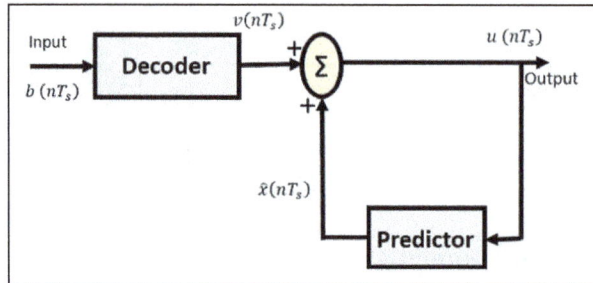

The notation of the signals is the same as the previous ones. In the absence of noise, the encoded receiver input will be the same as the encoded transmitter output.

The predictor assumes a value, based on the previous outputs. The input given to the decoder is processed and that output is summed up with the output of the predictor, to obtain a better output.

Digital Communication

The sampling rate of a signal should be higher than the Nyquist rate, to achieve better sampling. If this sampling interval in Differential PCM is reduced considerably, the sampleto-sample amplitude difference is very small, as if the difference is 1-bit quantization, then the step-size will be very small i.e., Δ (delta).

Delta Modulation

The type of modulation, where the sampling rate is much higher and in which the stepsize after quantization is of a smaller value Δ, such a modulation is termed as delta modulation.

Features of Delta Modulation

Following are some of the features of delta modulation:

- An over-sampled input is taken to make full use of the signal correlation.
- The quantization design is simple.
- The input sequence is much higher than the Nyquist rate.
- The quality is moderate.
- The design of the modulator and the demodulator is simple.
- The stair-case approximation of output waveform.

- The step-size is very small, i.e., Δ (delta).

- The bit rate can be decided by the user.

- This involves simpler implementation.

Delta Modulation is a simplified form of DPCM technique, also viewed as 1-bit DPCM scheme. As the sampling interval is reduced, the signal correlation will be higher.

Delta Modulator

The Delta Modulator comprises of a 1-bit quantizer and a delay circuit along with two summer circuits. Following is the block diagram of a delta modulator.

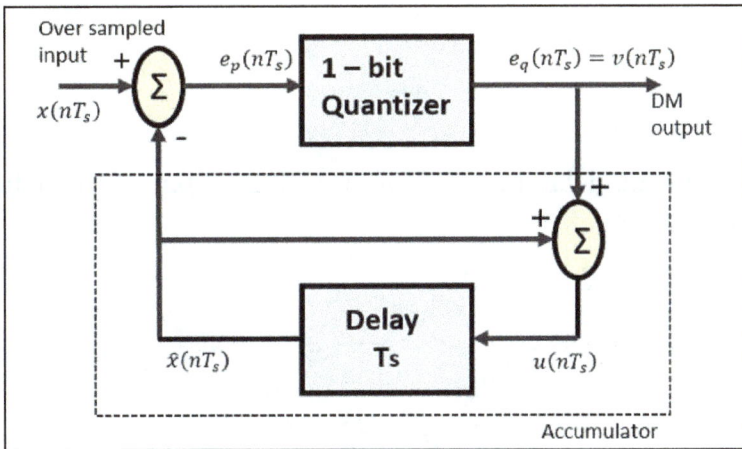

The predictor circuit in DPCM is replaced by a simple delay circuit in DM.

From the diagram, we have the notations as:

- $x(nT_s)$ = over sampled input,

- $e_p(nT_s)$ = summer output and quantizer input,

- $e_p(nT_s)$ = quantizer output = $v(nT_s)$,

- $\hat{x}(nT_s)$ = output of delay circuit,

- $u(nT_s)$ = input of delay circuit.

Using these notations, now we shall try to figure out the process of delta modulation.

$$e_p(nT_s) = x(nT_s) - \hat{x}(nT_s),$$

$$= x(nT_s) - u([n-1]T_s),$$

$$= x(nT_s) - \left[\left[\hat{x}[n-1]T_s\right] + v\left[[n-1]T_s\right]\right].$$

Further,

$$v(nT_s) = e_q(nT_s) = S.sig.\left[e_p(nT_s)\right],$$

$$u(nT_s) = \hat{x}(nT_s) + e_q(nT_s).$$

Where,

- $\hat{x}(nT_s)$ = the previous value of the delay circuit,
- $e_q(nT_s)$ = quantizer output = $v(nT_s)$.

Hence,

$$u(nT_s) = u\left([n-1]T_s\right) + v(nT_s)$$

which means,

The present input of the delay unit = (The previous output of the delay unit) + (the present quantizer output).

Assuming zero condition of Accumulation,

$$u(nT_s) = S\sum_{j=1}^{n} sig\left[e_q(jT_s)\right]$$

Accumulated version of DM output = $\sum_{j=1}^{n} v(jT_s)$.

Now, note that,

$$\hat{x}(nT_s) = u\left([n-1]T_s\right)$$

$$= \sum_{j=1}^{n-1} v(jT_s)$$

Delay unit output is an Accumulator output lagging by one sample.

A Stair-case approximated waveform will be the output of the delta modulator with the step-size as delta (Δ). The output quality of the waveform is moderate.

Delta Demodulator

The delta demodulator comprises of a low pass filter, a summer, and a delay circuit. The predictor circuit is eliminated here and hence no assumed input is given to the demodulator.

Following is the diagram for delta demodulator.

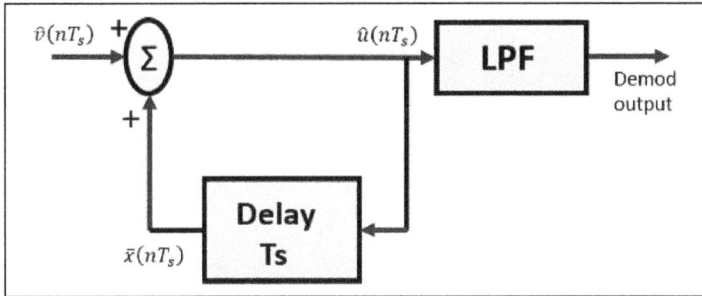

From the above diagram, we have the notations as:

- $\hat{v}(nT_s)$ is the input sample,

- $\hat{u}(nT_s)$ is the summer output,

- $\bar{x}(nT_s)$ is the delayed output.

A binary sequence will be given as an input to the demodulator. The stair-case approximated output is given to the LPF.

Low pass filter is used for many reasons, but the prominent reason is noise elimination for out-of-band signals. The step-size error that may occur at the transmitter is called granular noise, which is eliminated here. If there is no noise present, then the modulator output equals the demodulator input.

Advantages of DM over DPCM

- 1-bit quantizer.

- Very easy design of the modulator and the demodulator.

However, there exists some noise in DM.

- Slope Over load distortion (when Δ is small).

- Granular noise (when Δ is large).

Adaptive Delta Modulation

In digital modulation, we have come across certain problem of determining the step-size, which influences the quality of the output wave.

A larger step-size is needed in the steep slope of modulating signal and a smaller step-size is needed where the message has a small slope. The minute details get missed in the process. So, it would be better if we can control the adjustment of step-size, according to our requirement in order to obtain the sampling in a desired fashion. This is the concept of Adaptive Delta Modulation.

Following is the block diagram of Adaptive delta modulator.

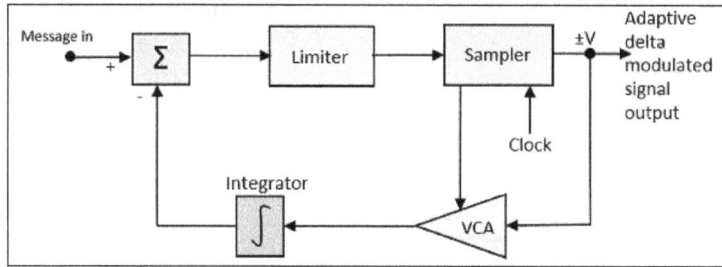

Adaptive Delta Modulation.

The gain of the voltage controlled amplifier is adjusted by the output signal from the sampler. The amplifier gain determines the step-size and both are proportional.

ADM quantizes the difference between the value of the current sample and the predicted value of the next sample. It uses a variable step height to predict the next values, for the faithful reproduction of the fast varying values.

Techniques in Digital Communication

There are a few techniques which have paved the basic path to digital communication processes. For the signals to get digitized, we have the sampling and quantizing techniques.

For them to be represented mathematically, we have LPC and digital multiplexing techniques.

Linear Predictive Coding

Linear Predictive Coding (LPC) is a tool which represents digital speech signals in linear predictive model. This is mostly used in audio signal processing, speech synthesis, speech recognition, etc.

Linear prediction is based on the idea that the current sample is based on the linear combination of past samples. The analysis estimates the values of a discrete-time signal as a linear function of the previous samples.

The spectral envelope is represented in a compressed form, using the information of the linear predictive model. This can be mathematically represented as:

$$s(n) = \sum_{k=1}^{p} \alpha_k s(n-k) \text{ for some value of p and } \alpha_k$$

Where,

- s(n) is the current speech sample,
- k is a particular sample,
- p is the most recent value,

- α_k is the predictor co-efficient,

- s(n - k) is the previous speech sample.

For LPC, the predictor co-efficient values are determined by minimizing the sum of squared differences (over a finite interval) between the actual speech samples and the linearly predicted ones.

This is a very useful method for encoding speech at a low bit rate. The LPC method is very close to the Fast Fourier Transform (FFT) method.

Multiplexing

Multiplexing is the process of combining multiple signals into one signal, over a shared medium. These signals, if analog in nature, the process is called as analog multiplexing. If digital signals are multiplexed, it is called as digital multiplexing.

The following figures represent MUX and DEMUX. Their primary use is in the field of communications.

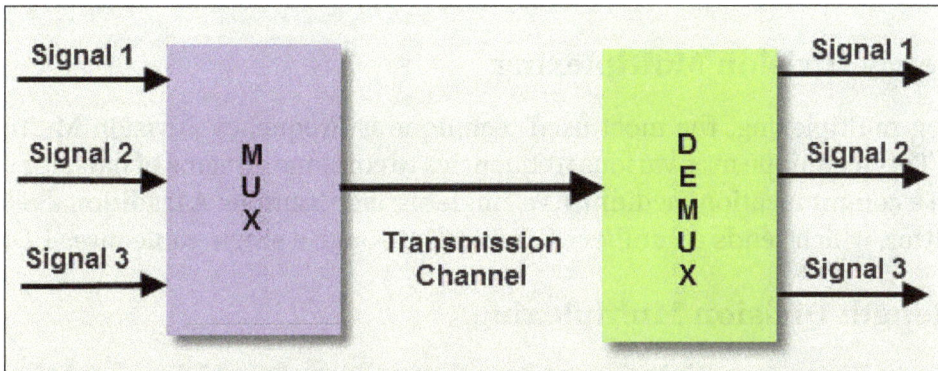

Multiplexing and Deultiplexing.

Multiplexing was first developed in telephony. A number of signals were combined to send through a single cable. The process of multiplexing divides a communication channel into several number of logical channels, allotting each one for a different message signal or a data stream to be transferred. The device that does multiplexing, can be called as a MUX. The reverse process, i.e., extracting the number of channels from one, which is done at the receiver is called as de-multiplexing. The device which does de-multiplexing is called as DEMUX.

Types of Multiplexers

There are mainly two types of multiplexers, namely analog and digital. They are further divided into FDM, WDM, and TDM. The following figure gives a detailed idea on this classification.

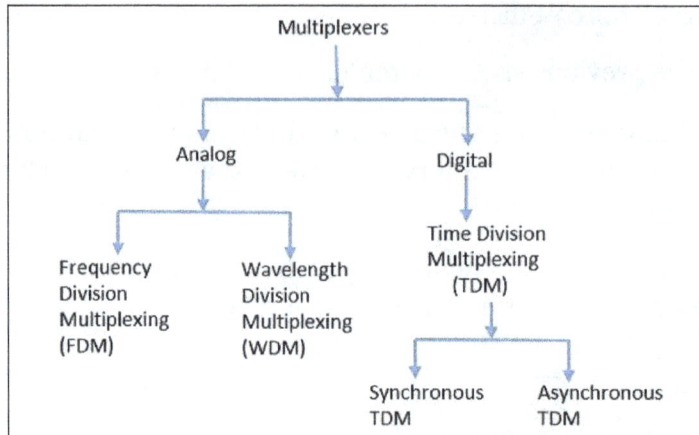

Analog Multiplexing

The analog multiplexing techniques involve signals which are analog in nature. The analog signals are multiplexed according to their frequency (FDM) or wavelength (WDM).

Frequency Division Multiplexing

In analog multiplexing, the most used technique is Frequency Division Multiplexing (FDM). This technique uses various frequencies to combine streams of data, for sending them on a communication medium, as a single signal. Example: A traditional television transmitter, which sends a number of channels through a single cable, uses FDM.

Wavelength Division Multiplexing

Wavelength Division multiplexing is an analog technique, in which many data streams of different wavelengths are transmitted in the light spectrum. If the wavelength increases, the frequency of the signal decreases. A prism which can turn different wavelengths into a single line, can be used at the output of MUX and input of DE-MUX. Example: Optical fiber communications use WDM technique to merge different wavelengths into a single light for communication.

Digital Multiplexing

The term digital represents the discrete bits of information. Hence, the available data is in the form of frames or packets, which are discrete.

Time Division Multiplexing

In TDM, the time frame is divided into slots. This technique is used to transmit a signal over a single communication channel, by allotting one slot for each message. Of all the types of TDM, the main ones are Synchronous and Asynchronous TDM.

Synchronous TDM

In Synchronous TDM, the input is connected to a frame. If there are 'n' number of connections, then the frame is divided into 'n' time slots. One slot is allocated for each input line. In this technique, the sampling rate is common to all signals and hence the same clock input is given. The MUX allocates the same slot to each device at all times.

Asynchronous TDM

In Asynchronous TDM, the sampling rate is different for each of the signals and a common clock is not required. If the allotted device, for a time-slot, transmits nothing and sits idle, then that slot is allotted to another device, unlike synchronous. This type of TDM is used in Asynchronous transfer mode networks.

Regenerative Repeater

For any communication system to be reliable, it should transmit and receive the signals effectively, without any loss. A PCM wave, after transmitting through a channel, gets distorted due to the noise introduced by the channel. The regenerative pulse compared with the original and received pulse, will be as shown in the following figure.

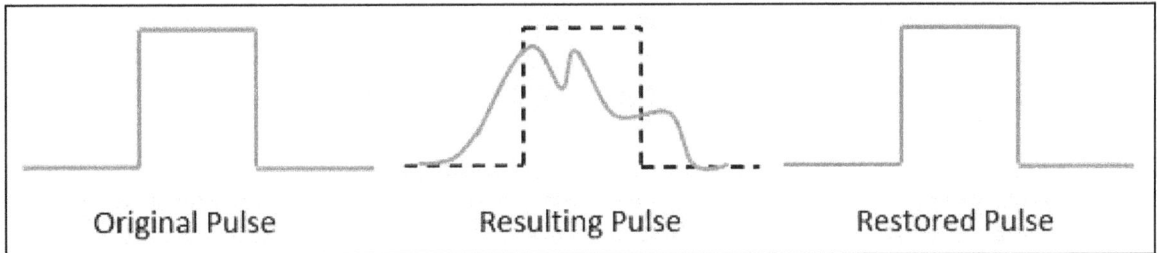

Original Pulse   Resulting Pulse   Restored Pulse

For a better reproduction of the signal, a circuit called as regenerative repeater is employed in the path before the receiver. This helps in restoring the signals from the losses occurred. Following is the diagrammatical representation.

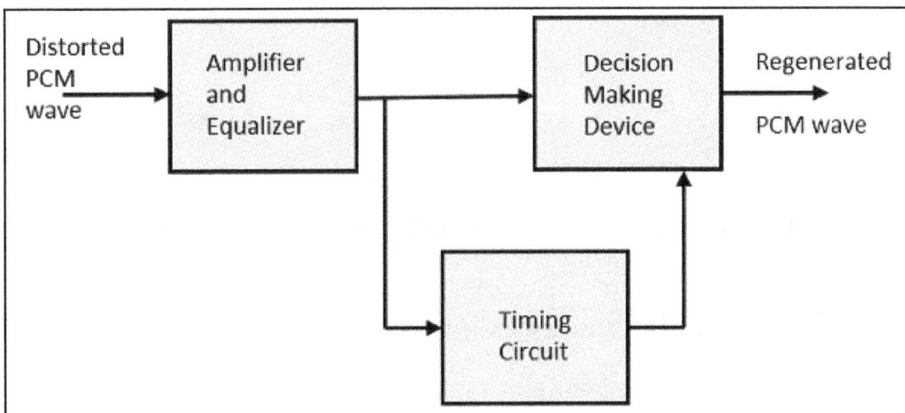

Block diagram of a regenerative repeater.

This consists of an equalizer along with an amplifier, a timing circuit, and a decision making device. Their working of each of the components is detailed as follows:

Equalizer

The channel produces amplitude and phase distortions to the signals. This is due to the transmission characteristics of the channel. The Equalizer circuit compensates these losses by shaping the received pulses.

Timing Circuit

To obtain a quality output, the sampling of the pulses should be done where the signal to noise ratio (SNR) is maximum. To achieve this perfect sampling, a periodic pulse train has to be derived from the received pulses, which is done by the timing circuit.

Hence, the timing circuit, allots the timing interval for sampling at high SNR, through the received pulses.

Decision Device

The timing circuit determines the sampling times. The decision device is enabled at these sampling times. The decision device decides its output based on whether the amplitude of the quantized pulse and the noise, exceeds a pre-determined value or not.

These are few of the techniques used in digital communications. There are other important techniques to be learned, called as data encoding techniques. Let us learn about them in the subsequent chapters, after taking a look at the line codes.

Digital Communication in Line Codes

A line code is the code used for data transmission of a digital signal over a transmission line. This process of coding is chosen so as to avoid overlap and distortion of signal such as inter-symbol interference.

Properties of Line Coding

Following are the properties of line coding:

- As the coding is done to make more bits transmit on a single signal, the bandwidth used is much reduced.

- For a given bandwidth, the power is efficiently used.

- The probability of error is much reduced.

- Error detection is done and the bipolar too has a correction capability.

- Power density is much favorable.

- The timing content is adequate.

- Long strings of 1s and 0s is avoided to maintain transparency.

Types of Line Coding

There are 3 types of Line Coding:

- Unipolar,

- Polar,

- Bi-polar.

Unipolar Signaling

Unipolar signaling is also called as On-off Keying or simply OOK. The presence of pulse represents a 1 and the absence of pulse represents a 0.

There are two variations in Unipolar signaling:

- Non Return to Zero (NRZ).

- Return to Zero (RZ).

Unipolar Non-return to Zero

In this type of unipolar signaling, a High in data is represented by a positive pulse called as Mark, which has a duration T_0 equal to the symbol bit duration. A Low in data input has no pulse.

The following figure clearly depicts this:

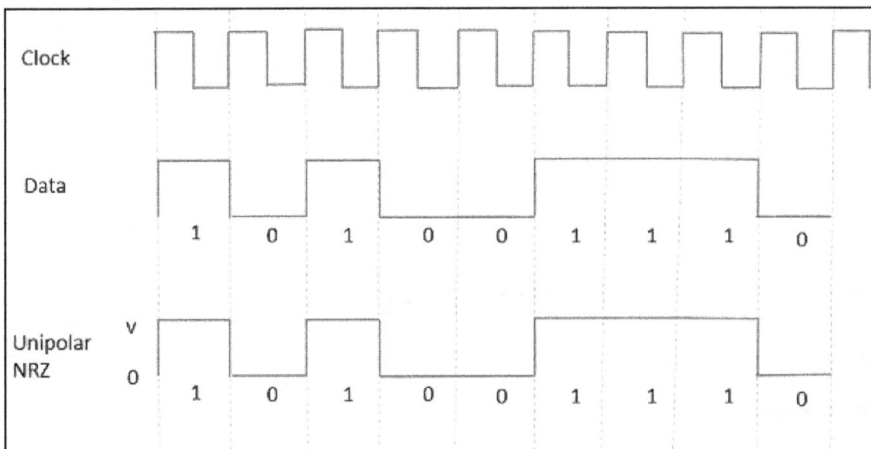

Advantages of Unipolar Non-return to Zero

The advantages of Unipolar NRZ are:

- It is simple.
- A lesser bandwidth is required.

Disadvantages of Unipolar Non-return to Zero

The disadvantages of Unipolar NRZ are:

- No error correction done.
- Presence of low frequency components may cause the signal droop.
- No clock is present.
- Loss of synchronization is likely to occur (especially for long strings of 1sand os).

Unipolar Return to Zero

In this type of unipolar signaling, a High in data, though represented by a Mark pulse, its duration T_0 is less than the symbol bit duration. Half of the bit duration remains high but it immediately returns to zero and shows the absence of pulse during the remaining half of the bit duration. It is clearly understood with the help of the following figure.

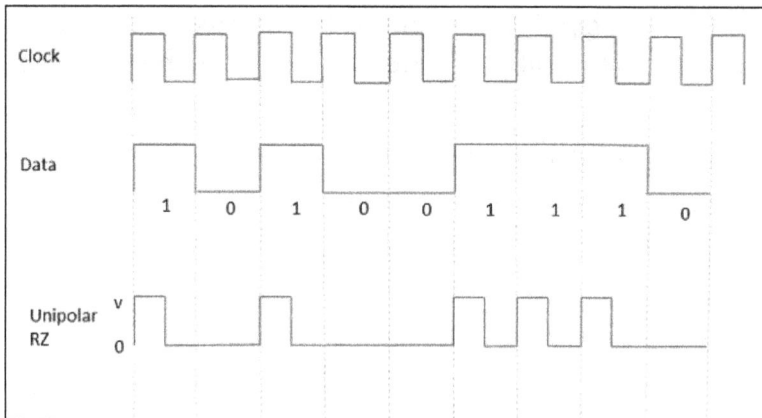

Advantages of Unipolar Return to Zero

The advantages of Unipolar RZ are:

- It is simple.
- The spectral line present at the symbol rate can be used as a clock.

Disadvantages of Unipolar Return to Zero

The disadvantages of Unipolar RZ are:

- No error correction.
- Occupies twice the bandwidth as unipolar NRZ.
- The signal droop is caused at the places where signal is non-zero at 0 Hz.

Polar Signaling

There are two methods of Polar Signaling:

- Polar NRZ.
- Polar RZ.

Polar NRZ

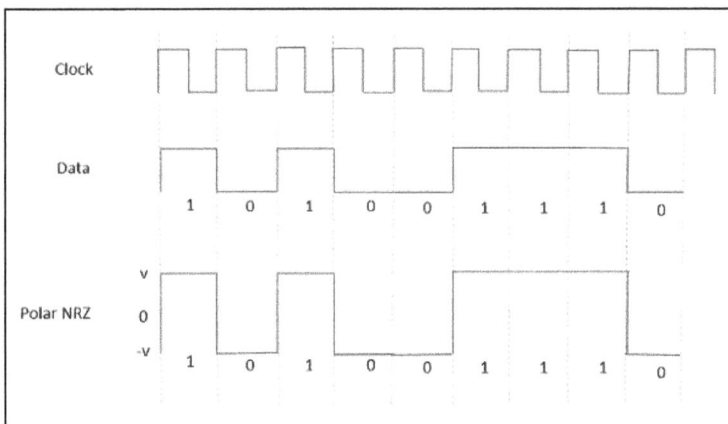

In this type of Polar signaling, a High in data is represented by a positive pulse, while a Low in data is represented by a negative pulse. The following figure depicts this well.

Advantages of Polar NRZ

The advantages of Polar NRZ are:

- It is simple.
- No low-frequency components are present.

Disadvantages of Polar NRZ

The disadvantages of Polar NRZ are:

- No error correction.

- No clock is present.
- The signal droop is caused at the places where the signal is non-zero at 0 Hz.

Polar RZ

In this type of Polar signaling, a High in data, though represented by a Mark pulse, its duration T_0 is less than the symbol bit duration. Half of the bit duration remains high but it immediately returns to zero and shows the absence of pulse during the remaining half of the bit duration.

However, for a Low input, a negative pulse represents the data, and the zero level remains same for the other half of the bit duration. The following figure depicts this clearly.

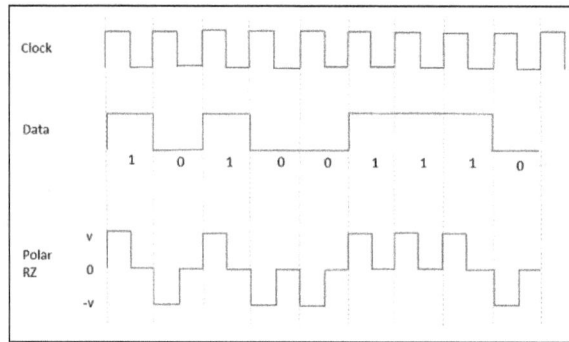

Advantages of Polar RZ

The advantages of Polar RZ are:

- It is simple.
- No low-frequency components are present.

Disadvantages of Polar RZ

The disadvantages of Polar RZ are:

- No error correction.
- No clock is present.
- Occupies twice the bandwidth of Polar NRZ.
- The signal droop is caused at places where the signal is non-zero at 0 Hz.

Bipolar Signaling

This is an encoding technique which has three voltage levels namely +, - and 0. Such a signal is called as duo-binary signal.

An example of this type is Alternate Mark Inversion (AMI). For a 1, the voltage level gets a transition from + to − or from − to +, having alternate 1s to be of equal polarity. A 0 will have a zero voltage level.

Even in this method, we have two types:

- Bipolar NRZ,

- Bipolar RZ

From the models we have learnt the difference between NRZ and RZ. It just goes in the same way here too. The following figure clearly depicts this.

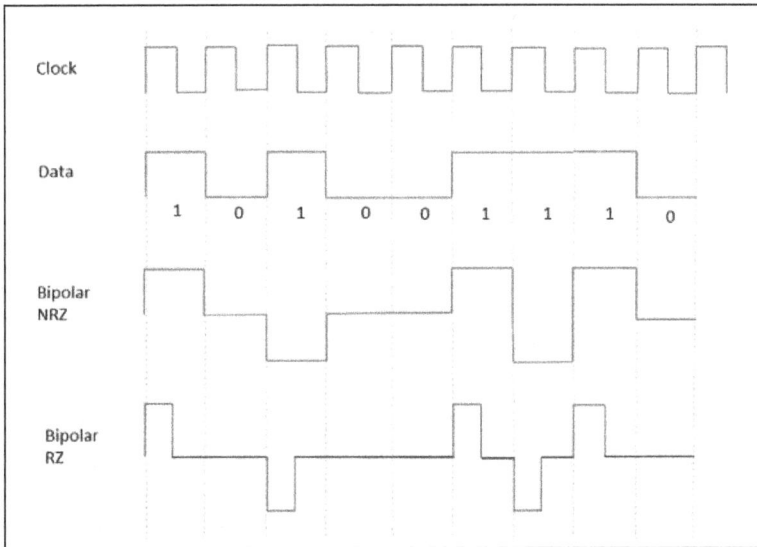

The above figure has both the Bipolar NRZ and RZ waveforms. The pulse duration and symbol bit duration are equal in NRZ type, while the pulse duration is half of the symbol bit duration in RZ type.

Advantages of Bipolar Signaling

Following are the advantages:

- It is simple.

- No low-frequency components are present.

- Occupies low bandwidth than unipolar and polar NRZ schemes.

- This technique is suitable for transmission over AC coupled lines, as signal drooping doesn't occur here.

- A single error detection capability is present in this.

Disadvantages of Bipolar Signaling

Following are the disadvantages:

- No clock is present.

- Long strings of data cause loss of synchronization.

Power Spectral Density

The function which describes how the power of a signal got distributed at various frequencies, in the frequency domain is called as Power Spectral Density (PSD).

PSD is the Fourier Transform of Auto-correlation (Similarity between observations). It is in the form of a rectangular pulse.

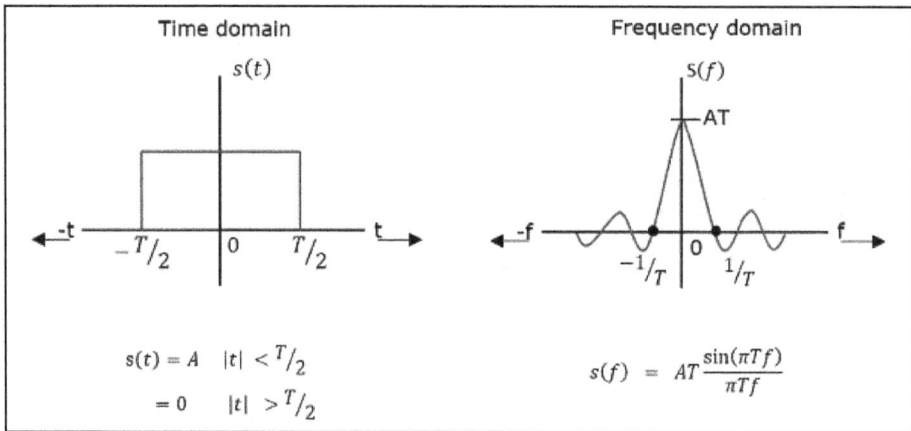

PSD Derivation

According to the Einstein-Wiener-Khintchine theorem, if the auto correlation function or power spectral density of a random process is known, the other can be found exactly.

Hence, to derive the power spectral density, we shall use the time auto-correlation $(R_x(\tau))$ of a power signal $x(t)$ as shown below:

$$R_x(\tau) = \lim_{T_p \to \infty} \frac{1}{T_p} \int_{\frac{-T_p}{2}}^{\frac{T_p}{2}} x(t)x(t+\tau)\,dt$$

Since $x(t)$ consists of impulses, $R_x(\tau)$ can be written as,

$$R_x(\tau) = \frac{1}{T} \sum_{n=-\infty}^{R_x(\tau)} R_n \delta(\tau - nT)$$

Where, $R_n = \lim_{N \to \infty} \frac{1}{N} \sum_k a_k a_{k+n}.$

Getting to know that $R_n = R_{-n}$ for real signals, we have,

$$S_x(w) = -\left(R + 2\sum_{n}^{\infty} R_n \cos nwT \right)$$

Since the pulse filter has the spectrum of $(w) \leftrightarrow f(t)$, we have,

$$s_y(w) = |F(w)|^2 S_x(w)$$

$$= \frac{|F(w)|^2}{T} \left(\sum_{n=-\infty}^{\infty} R_n e^{-jnwT_b} \right)$$

$$= \frac{|F(w)|^2}{T} \left(R_0 + 2\sum_{n=1}^{\infty} R_n \cos nwT \right)$$

Hence, we get the equation for Power Spectral Density. Using this, we can find the PSD of various line codes.

Data Encoding Techniques

Encoding is the process of converting the data or a given sequence of characters, symbols, alphabets etc., into a specified format, for the secured transmission of data. Decoding is the reverse process of encoding which is to extract the information from the converted format.

Data Encoding

Encoding is the process of using various patterns of voltage or current levels to represent 1s and 0s of the digital signals on the transmission link. The common types of line encoding are Unipolar, Polar, Bipolar, and Manchester.

Encoding Techniques

The data encoding technique is divided into the following types, depending upon the type of data conversion:

- Analog data to Analog signals: The modulation techniques such as Amplitude Modulation, Frequency Modulation and Phase Modulation of analog signals, fall under this category.

- Analog data to Digital signals: This process can be termed as digitization, which is done by Pulse Code Modulation (PCM). Hence, it is nothing but digital modulation. As we have already discussed, sampling and quantization are the important factors in this. Delta Modulation gives a better output than PCM.

- Digital data to Analog signals: The modulation techniques such as Amplitude Shift Keying (ASK), Frequency Shift Keying (FSK), Phase Shift Keying (PSK), etc., fall under this category.

- Digital data to Digital signals: There are several ways to map digital data to digital signals.

Non-return to Zero

NRZ Codes has 1 for High voltage level and 0 for Low voltage level. The main behavior of NRZ codes is that the voltage level remains constant during bit interval. The end or start of a bit will not be indicated and it will maintain the same voltage state, if the value of the previous bit and the value of the present bit are same. The following figure explains the concept of NRZ coding.

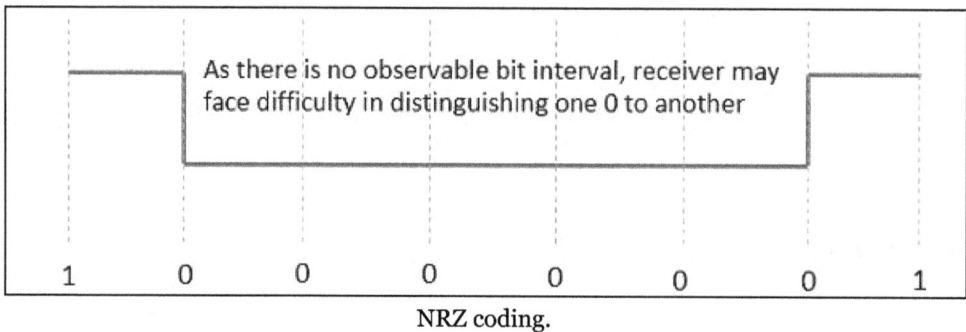

As there is no observable bit interval, receiver may face difficulty in distinguishing one 0 to another

| 1 | 0 | 0 | 0 | 0 | 0 | 0 | 1 |

NRZ coding.

If the above example is considered, as there is a long sequence of constant voltage level and the clock synchronization may be lost due to the absence of bit interval, it becomes difficult for the receiver to differentiate between 0 and 1. There are two variations in NRZ namely:

NRZ–Level

There is a change in the polarity of the signal, only when the incoming signal changes from 1 to 0 or from 0 to 1. It is the same as NRZ, however, the first bit of the input signal should have a change of polarity.

NRZ–Inverted

If a 1 occurs at the incoming signal, then there occurs a transition at the beginning of the bit interval. For a 0 at the incoming signal, there is no transition at the beginning of the bit interval.

NRZ codes has a disadvantage that the synchronization of the transmitter clock with the receiver clock gets completely disturbed, when there is a string of 1s and 0s. Hence, a separate clock line needs to be provided.

Bi-phase Encoding

The signal level is checked twice for every bit time, both initially and in the middle. Hence, the clock rate is double the data transfer rate and thus the modulation rate is also doubled. The clock is taken from the signal itself. The bandwidth required for this coding is greater.

There are two types of Bi-phase Encoding:

- Bi-phase Manchester.
- Differential Manchester.

Bi-phase Manchester

In this type of coding, the transition is done at the middle of the bit-interval. The transition for the resultant pulse is from High to Low in the middle of the interval, for the input bit 1. While the transition is from Low to High for the input bit 0.

Differential Manchester

In this type of coding, there always occurs a transition in the middle of the bit interval. If there occurs a transition at the beginning of the bit interval, then the input bit is 0. If no transition occurs at the beginning of the bit interval, then the input bit is 1.

The following figure illustrates the waveforms of NRZ-L, NRZ-I, Bi-phase Manchester and Differential Manchester coding for different digital inputs.

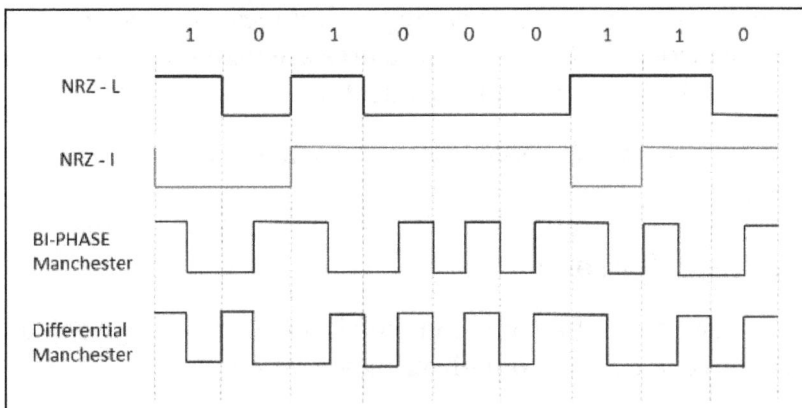

Block Coding

Among the types of block coding, the famous ones are 4B/5B encoding and 8B/6T encoding. The number of bits are processed in different manners, in both of these processes.

4B/5B Encoding

In Manchester encoding, to send the data, the clocks with double speed is required rather than NRZ coding. Here, as the name implies, 4 bits of code is mapped with 5 bits, with a minimum number of 1 bits in the group.

The clock synchronization problem in NRZ-I encoding is avoided by assigning an equivalent word of 5 bits in the place of each block of 4 consecutive bits. These 5-bit words are predetermined in a dictionary.

The basic idea of selecting a 5-bit code is that, it should have one leading 0 and it should have no more than two trailing 0s. Hence, these words are chosen such that two transactions take place per block of bits.

8B/6T Encoding

We have used two voltage levels to send a single bit over a single signal. But if we use more than 3 voltage levels, we can send more bits per signal.

For example, if 6 voltage levels are used to represent 8 bits on a single signal, then such encoding is termed as 8B/6T encoding. Hence in this method, we have as many as 729 combinations for signal and 256 combinations for bits.

These are the techniques mostly used for converting digital data into digital signals by compressing or coding them for reliable transmission of data.

Pulse Shaping

After going through different types of coding techniques, we have an idea on how the data is prone to distortion and how the measures are taken to prevent it from getting affected so as to establish a reliable communication.

There is another important distortion which is most likely to occur, called as Inter-symbol Interference (ISI).

Inter-symbol Interference

This is a form of distortion of a signal, in which one or more symbols interfere with subsequent signals, causing noise or delivering a poor output.

Causes of ISI

The main causes of ISI are:

- Multi-path Propagation.
- Non-linear frequency in channels.

The ISI is unwanted and should be completely eliminated to get a clean output. The causes of ISI should also be resolved in order to lessen its effect.

To view ISI in a mathematical form present in the receiver output, we can consider the receiver output.

The receiving filter output $y(t)$ is sampled at time $t_i = iT_b$ (with i taking on integer values), yielding:

$$y(t_i) = \mu \sum_{k=-\infty}^{\infty} a_k p(iT_b - kT_b - kT_b)$$

$$= \mu a_i + \mu \sum_{\substack{k=-\infty \\ k \neq i}}^{\infty} a_k p(iT_b - kT_b)$$

In the above equation, the first term μa_i is produced by the ith transmitted bit.

The second term represents the residual effect of all other transmitted bits on the decoding of the ith bit. This residual effect is called as Inter Symbol Interference.

In the absence of ISI, the output will be:

$$y(t_i) = \mu a_i$$

This equation shows that the ith bit transmitted is correctly reproduced. However, the presence of ISI introduces bit errors and distortions in the output.

While designing the transmitter or a receiver, it is important that you minimize the effects of ISI, so as to receive the output with the least possible error rate.

Correlative Coding

So far, we've discussed that ISI is an unwanted phenomenon and degrades the signal. But the same ISI if used in a controlled manner, is possible to achieve a bit rate of 2W bits per second in a channel of bandwidth W Hertz. Such a scheme is called as Correlative Coding or Partial response signaling schemes.

Since the amount of ISI is known, it is easy to design the receiver according to the requirement so as to avoid the effect of ISI on the signal. The basic idea of correlative coding is achieved by considering an example of Duo-binary Signaling.

Duo-binary Signaling

The name duo-binary means doubling the binary system's transmission capability. To understand this, let us consider a binary input sequence $\{a_k\}$ consisting of uncorrelated binary digits each having a duration T_a seconds. In this, the signal 1 is represented by a +1 volt and the symbol 0 by a -1 volt.

Therefore, the duo-binary coder output c_k is given as the sum of present binary digit a_k and the previous value a_{k-1} as shown in the following equation.

$$c_k = a_k + a_{k-1}$$

The above equation states that the input sequence of uncorrelated binary sequence $\{a_k\}$ is changed into a sequence of correlated three level pulses $\{c_k\}$. This correlation between the pulses may be understood as introducing ISI in the transmitted signal in an artificial manner.

Eye Pattern

An effective way to study the effects of ISI is the Eye Pattern. The name Eye Pattern was given from its resemblance to the human eye for binary waves. The interior region of the eye pattern is called the eye opening. The following figure shows the image of an eye-pattern.

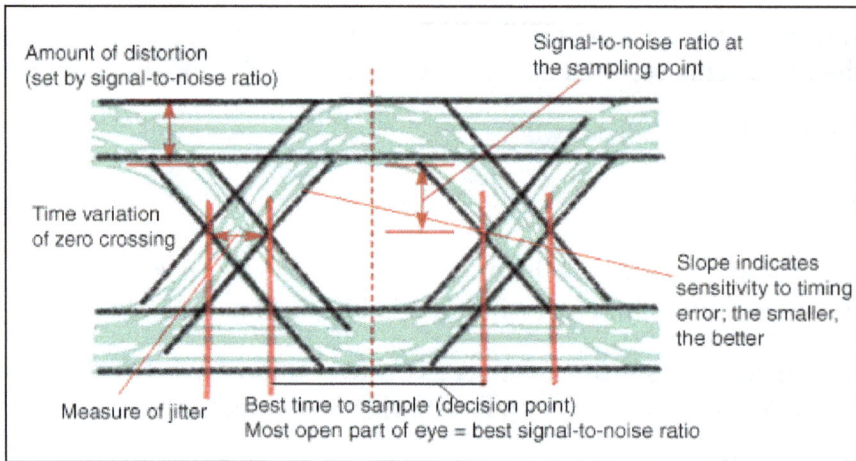

Jitter is the short-term variation of the instant of digital signal, from its ideal position, which may lead to data errors.

When the effect of ISI increases, traces from the upper portion to the lower portion of the eye opening increases and the eye gets completely closed, if ISI is very high.

An eye pattern provides the following information about a particular system:

- Actual eye patterns are used to estimate the bit error rate and the signal-to-noise ratio.

- The width of the eye opening defines the time interval over which the received wave can be sampled without error from ISI.

- The instant of time when the eye opening is wide, will be the preferred time for sampling.

- The rate of the closure of the eye, according to the sampling time, determines how sensitive the system is to the timing error.

- The height of the eye opening, at a specified sampling time, defines the margin over noise.

Hence, the interpretation of eye pattern is an important consideration.

Equalization

For reliable communication to be established, we need to have a quality output. The transmission losses of the channel and other factors affecting the quality of the signal, have to be treated. The most occurring loss, as we have discussed, is the ISI.

To make the signal free from ISI, and to ensure a maximum signal to noise ratio, we need to implement a method called Equalization. The following figure shows an equalizer in the receiver portion of the communication system.

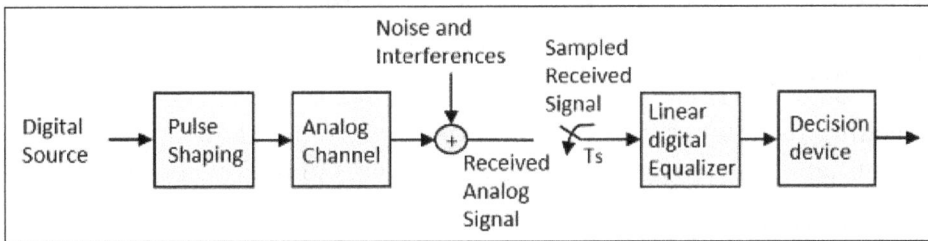

The noise and interferences which are denoted in the figure are likely to occur, during transmission. The regenerative repeater has an equalizer circuit, which compensates the transmission losses by shaping the circuit. The Equalizer is feasible to get implemented.

Error Probability and Figure-of-merit

The rate at which data can be communicated is called the data rate. The rate at which error occurs in the bits, while transmitting data is called the Bit Error Rate (BER). The probability of the occurrence of BER is the Error Probability. The increase in Signal to Noise Ratio (SNR) decreases the BER, hence the Error Probability also gets decreased.

In an Analog receiver, the figure of merit at the detection process can be termed as the ratio of output SNR to the input SNR. A greater value of figure-of-merit will be an advantage.

Digital Modulation Techniques

These techniques are also called as Digital Modulation techniques. Digital Modulation provides more information capacity, high data security, quicker system availability with great quality communication. Hence, digital modulation techniques have a greater demand, for their capacity to convey larger amounts of data than analog modulation techniques.

There are many types of digital modulation techniques and also their combinations, depending upon the need.

- Amplitude Shift Keying: The amplitude of the resultant output depends upon the input data whether it should be a zero level or a variation of positive and negative, depending upon the carrier frequency.

- Frequency Shift Keying: The frequency of the output signal will be either high or low, depending upon the input data applied.

- Phase Shift Keying: The phase of the output signal gets shifted depending upon the input. These are mainly of two types, namely Binary Phase Shift Keying (BPSK) and Quadrature Phase Shift Keying (QPSK), according to the number of phase shifts. The other one is Differential Phase Shift Keying (DPSK) which changes the phase according to the previous value.

M-ary Encoding

M-ary Encoding techniques are the methods where more than two bits are made to transmit simultaneously on a single signal. This helps in the reduction of bandwidth.

The types of M-ary techniques are:

- M-ary ASK.

- M-ary FSK.

- M-ary PSK.

Amplitude Shift Keying

Amplitude Shift Keying (ASK) is a type of Amplitude Modulation which represents the binary data in the form of variations in the amplitude of a signal.

The following figure represents ASK modulated waveform along with its input.

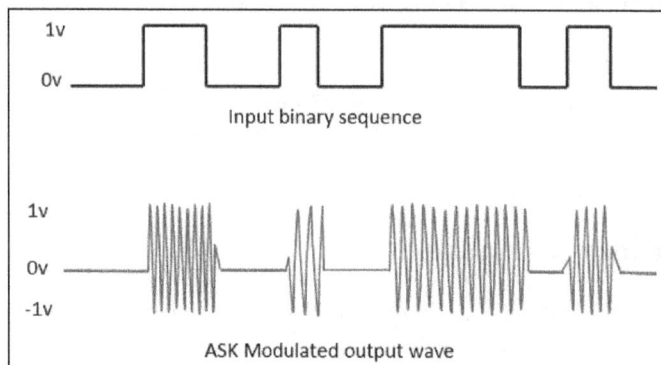

Any modulated signal has a high frequency carrier. The binary signal when ASK modulated, gives a zero value for Low input while it gives the carrier output for High input.

To find the process of obtaining this ASK modulated wave, let us learn about the working of the ASK modulator.

ASK Modulator

The ASK modulator block diagram comprises of the carrier signal generator, the binary sequence from the message signal and the band-limited filter. Following is the block diagram of the ASK Modulator.

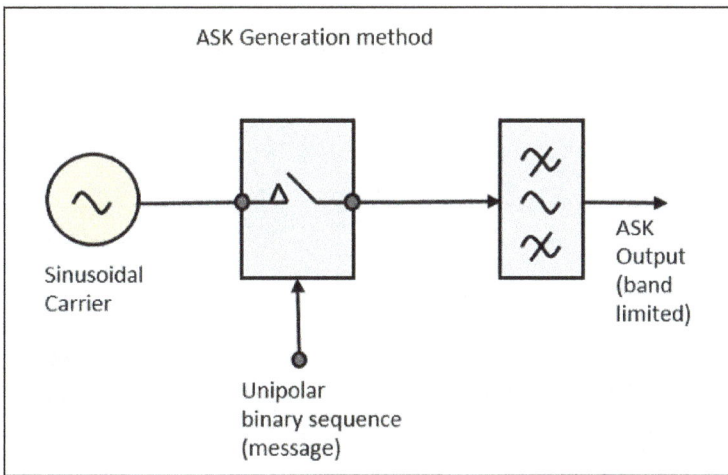

The carrier generator, sends a continuous high-frequency carrier. The binary sequence from the message signal makes the unipolar input to be either High or Low. The high signal closes the switch, allowing a carrier wave. Hence, the output will be the carrier signal at high input. When there is low input, the switch opens, allowing no voltage to appear. Hence, the output will be low.

The band-limiting filter, shapes the pulse depending upon the amplitude and phase characteristics of the band-limiting filter or the pulse-shaping filter.

ASK Demodulator

There are two types of ASK Demodulation techniques. They are:

• Asynchronous ASK Demodulation/detection.

• Synchronous ASK Demodulation/detection.

The clock frequency at the transmitter when matches with the clock frequency at the receiver, it is known as a Synchronous method, as the frequency gets synchronized. Otherwise, it is known as Asynchronous.

Asynchronous ASK Demodulator

The Asynchronous ASK detector consists of a half-wave rectifier, a low pass filter, and a comparator.

The modulated ASK signal is given to the half-wave rectifier, which delivers a positive half output. The low pass filter suppresses the higher frequencies and gives an envelope detected output from which the comparator delivers a digital output.

Following is the block diagram for the same:

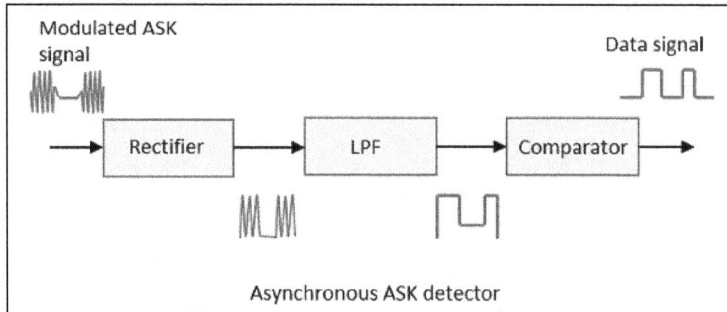

Asynchronous ASK detector

Synchronous ASK Demodulator

Synchronous ASK detector consists of a Square law detector, low pass filter, a comparator, and a voltage limiter. Following is the block diagram for the same.

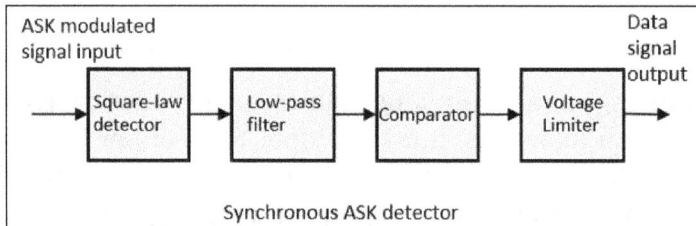

Synchronous ASK detector

The ASK modulated input signal is given to the Square law detector. A square law detector is one whose output voltage is proportional to the square of the amplitude modulated input voltage. The low pass filter minimizes the higher frequencies. The comparator and the voltage limiter help to get a clean digital output.

Frequency Shift Keying

Frequency Shift Keying (FSK) is the digital modulation technique in which the frequency of the carrier signal varies according to the digital signal changes. FSK is a scheme of frequency modulation.

The output of a FSK modulated wave is high in frequency for a binary High input and is low in frequency for a binary Low input. The binary 1s and 0s are called Mark and Space frequencies.

The following image is the diagrammatic representation of FSK modulated waveform along with its input.

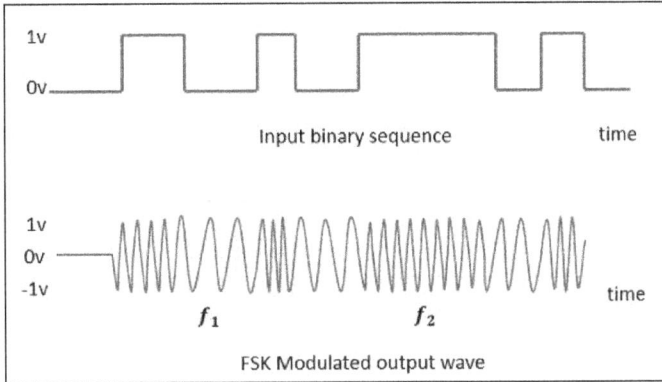

FSK Modulated output wave

To find the process of obtaining this FSK modulated wave, let us know about the working of a FSK modulator.

FSK Modulator

The FSK modulator block diagram comprises of two oscillators with a clock and the input binary sequence. Following is its block diagram.

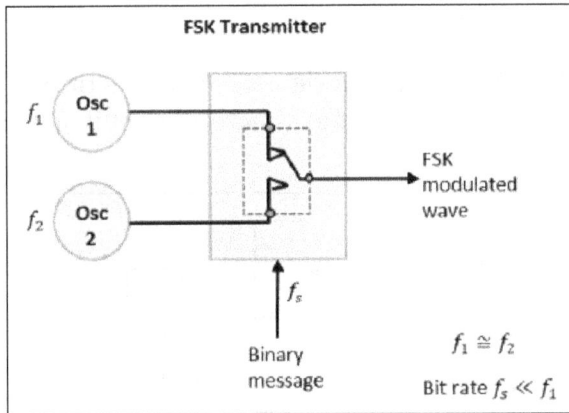

The two oscillators, producing a higher and a lower frequency signals, are connected to a switch along with an internal clock. To avoid the abrupt phase discontinuities of the output waveform during the transmission of the message, a clock is applied to both the oscillators, internally. The binary input sequence is applied to the transmitter so as to choose the frequencies according to the binary input.

FSK Demodulator

There are different methods for demodulating a FSK wave. The main methods of FSK detection are asynchronous detector and synchronous detector. The synchronous detector is a coherent one, while asynchronous detector is a non-coherent one.

Asynchronous FSK Detector

The block diagram of Asynchronous FSK detector consists of two band pass filters, two envelope detectors, and a decision circuit. Following is the diagrammatic representation.

The FSK signal is passed through the two Band Pass Filters (BPFs), tuned to Space and Mark frequencies. The output from these two BPFs look like ASK signal, which is given to the envelope detector. The signal in each envelope detector is modulated asynchronously.

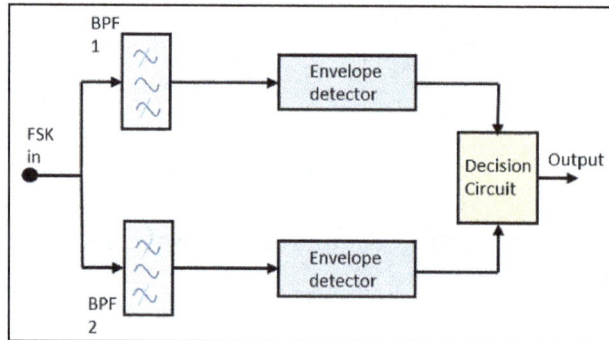

The decision circuit chooses which output is more likely and selects it from any one of the envelope detectors. It also re-shapes the waveform to a rectangular one.

Synchronous FSK Detector

The block diagram of Synchronous FSK detector consists of two mixers with local oscillator circuits, two band pass filters and a decision circuit. Following is the diagrammatic representation.

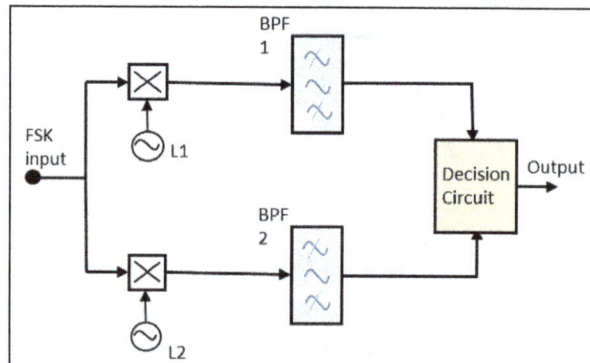

The FSK signal input is given to the two mixers with local oscillator circuits. These two are connected to two band pass filters. These combinations act as demodulators and the decision circuit chooses which output is more likely and selects it from any one of the detectors. The two signals have a minimum frequency separation.

For both of the demodulators, the bandwidth of each of them depends on their bit rate. This synchronous demodulator is a bit complex than asynchronous type demodulators.

Phase Shift Keying

Phase Shift Keying (PSK) is the digital modulation technique in which the phase of the carrier signal is changed by varying the sine and cosine inputs at a particular time. PSK technique is widely used for wireless LANs, bio-metric, contactless operations, along with RFID and Bluetooth communications. PSK is of two types, depending upon the phases the signal gets shifted.

Binary Phase Shift Keying

This is also called as 2-phase PSK or Phase Reversal Keying. In this technique, the sine wave carrier takes two phase reversals such as 0° and 180°.

BPSK is basically a Double Side Band Suppressed Carrier (DSBSC) modulation scheme, for message being the digital information.

Quadrature Phase Shift Keying

This is the phase shift keying technique, in which the sine wave carrier takes four phase reversals such as 0°, 90°, 180°, and 270°.

If this kind of techniques are further extended, PSK can be done by eight or sixteen values also, depending upon the requirement.

BPSK Modulator

The block diagram of Binary Phase Shift Keying consists of the balance modulator which has the carrier sine wave as one input and the binary sequence as the other input. Following is the diagrammatic representation.

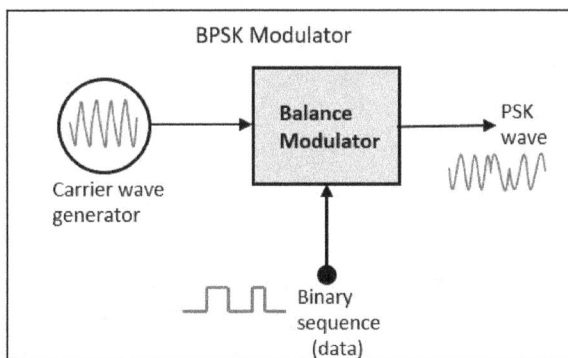

The modulation of BPSK is done using a balance modulator, which multiplies the two signals applied at the input. For a zero binary input, the phase will be 0° and for a high input, the phase reversal is of 180°.

Following is the diagrammatic representation of BPSK Modulated output wave along with its given input.

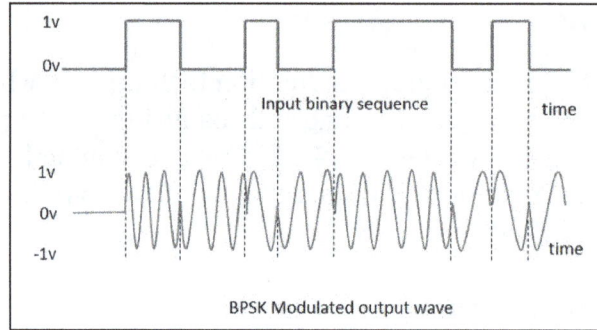

The output sine wave of the modulator will be the direct input carrier or the inverted (180° phase shifted) input carrier, which is a function of the data signal.

BPSK Demodulator

The block diagram of BPSK demodulator consists of a mixer with local oscillator circuit, a band pass filter, a two-input detector circuit. The diagram is as follows.

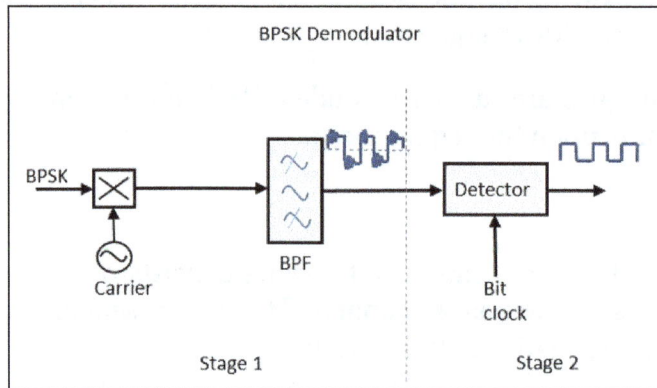

By recovering the band-limited message signal, with the help of the mixer circuit and the band pass filter, the first stage of demodulation gets completed. The base band signal which is band limited is obtained and this signal is used to regenerate the binary message bit stream.

In the next stage of demodulation, the bit clock rate is needed at the detector circuit to produce the original binary message signal. If the bit rate is a sub-multiple of the carrier frequency, then the bit clock regeneration is simplified. To make the circuit easily understandable, a decision-making circuit may also be inserted at the 2nd stage of detection.

Quadrature Phase Shift Keying

The Quadrature Phase Shift Keying (QPSK) is a variation of BPSK, and it is also a Double Side Band Suppressed Carrier (DSBSC) modulation scheme, which sends two bits of digital information at a time, called as bigits.

Instead of the conversion of digital bits into a series of digital stream, it converts them into bit pairs. This decreases the data bit rate to half, which allows space for the other users.

QPSK Modulator

The QPSK Modulator uses a bit-splitter, two multipliers with local oscillator, a 2-bit serial to parallel converter, and a summer circuit. Following is the block diagram for the same.

At the modulator's input, the message signal's even bits (i.e., 2nd bit, 4th bit, 6th bit, etc.) and odd bits (i.e., 1st bit, 3rd bit, 5th bit, etc.) are separated by the bits splitter and are multiplied with the same carrier to generate odd BPSK (called as PSK_I) and even BPSK (called as PSK_Q). The PSK_Q signal is anyhow phase shifted by 90° before being modulated.

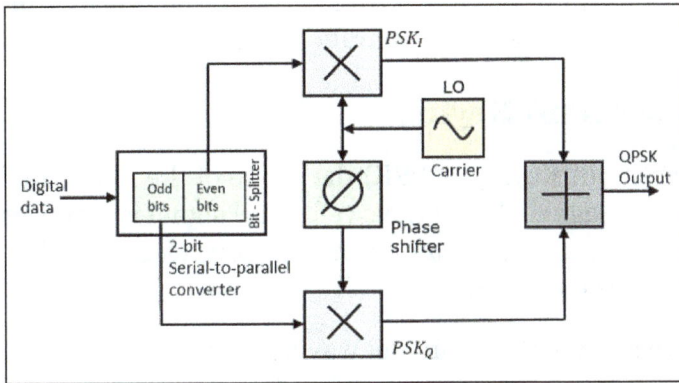

The QPSK waveform for two-bits input is as follows, which shows the modulated result for different instances of binary inputs.

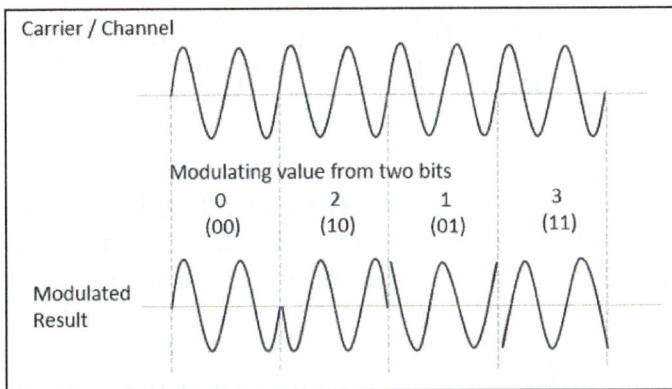

QPSK Demodulator

The QPSK Demodulator uses two product demodulator circuits with local oscillator, two band pass filters, two integrator circuits, and a 2-bit parallel to serial converter. Following is the diagram for the same.

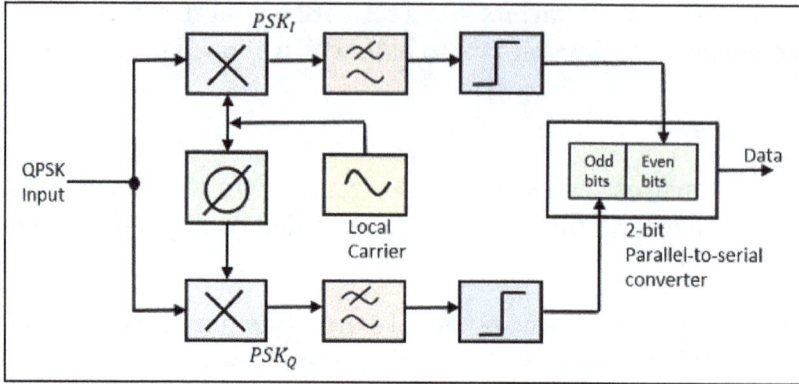

The two product detectors at the input of demodulator simultaneously demodulate the two BPSK signals. The pair of bits are recovered here from the original data. These signals after processing, are passed to the parallel to serial converter.

Differential Phase Shift Keying

In Differential Phase Shift Keying (DPSK) the phase of the modulated signal is shifted relative to the previous signal element. No reference signal is considered here. The signal phase follows the high or low state of the previous element. This DPSK technique doesn't need a reference oscillator.

The following figure represents the model waveform of DPSK.

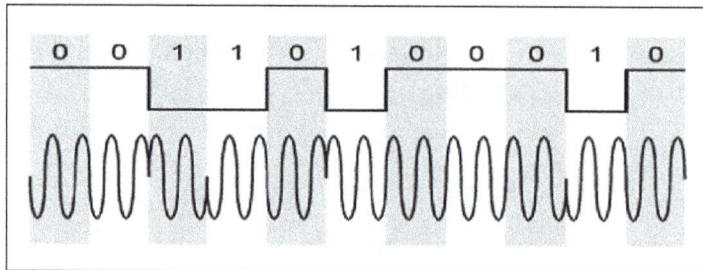

It is seen from the above figure that, if the data bit is Low i.e., 0, then the phase of the signal is not reversed, but continued as it was. If the data is a High i.e., 1, then the phase of the signal is reversed, as with NRZI, invert on 1 (a form of differential encoding).

If we observe the above waveform, we can say that the High state represents an Min the modulating signal and the Low state represents a W in the modulating signal.

DPSK Modulator

DPSK is a technique of BPSK, in which there is no reference phase signal. Here, the transmitted signal itself can be used as a reference signal. Following is the diagram of DPSK Modulator.

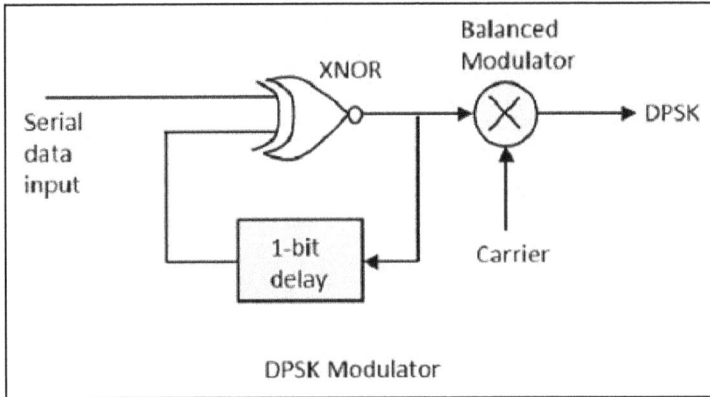

DPSK Modulator

DPSK encodes two distinct signals, i.e., the carrier and the modulating signal with 180° phase shift each. The serial data input is given to the XNOR gate and the output is again fed back to the other input through 1-bit delay. The output of the XNOR gate along with the carrier signal is given to the balance modulator, to produce the DPSK modulated signal.

DPSK Demodulator

In DPSK demodulator, the phase of the reversed bit is compared with the phase of the previous bit. Following is the block diagram of DPSK demodulator.

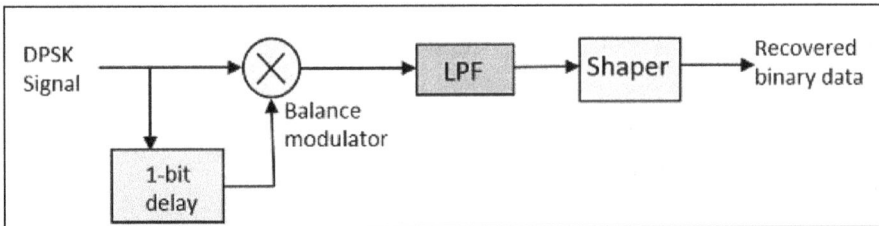

From the figure, it is evident that the balance modulator is given the DPSK signal along with 1-bit delay input. That signal is made to confine to lower frequencies with the help of LPF. Then it is passed to a shaper circuit, which is a comparator or a Schmitt trigger circuit, to recover the original binary data as the output.

Entropy

When we observe the possibilities of the occurrence of an event, how surprising or uncertain it would be, it means that we are trying to have an idea on the average content of the information from the source of the event.

Entropy can be defined as a measure of the average information content per source symbol. Claude Shannon, the "father of the Information Theory", provided a formula for it as:

$$H = -\sum_i p_i \log_b p_i$$

Where p_i is the probability of the occurrence of character number i from a given stream of characters and b is the base of the algorithm used. Hence, this is also called as Shannon's Entropy.

The amount of uncertainty remaining about the channel input after observing the channel output, is called as Conditional Entropy. It is denoted by $H(x \mid y)$.

Mutual Information

Let us consider a channel whose output is Y and input is X.

Let the entropy for prior uncertainty be X = H(x).

To know about the uncertainty of the output, after the input is applied, let us consider Conditional Entropy, given that $Y = y_k$.

$$H(x \mid yk) = \sum_{j=0}^{j-1} p(x_j \mid yk) \log_2 \left[\frac{1}{p(x_j \mid yk)} \right]$$

This is a random variable for $H(X \mid y = y_0)$... ... ... $H(X \mid y = y_k)$ with probabilities $p(y_0)$... ... ... $p(y_{k-1})$ respectively.

The mean value of $H(X \mid y = y_k)$ for output alphabet y is:

$$H(X \mid Y) = \sum_{k=0}^{k-1} H(X \mid yy = y_k) p(y_k)$$

$$= \sum_{k=0}^{k-1} \sum_{j=0}^{j-1} p(x_j \mid y_k) p(y_k) \log_2 \left[\frac{1}{p(x_j \mid y_k)} \right]$$

$$= \sum_{k=0}^{k-1} \sum_{j=0}^{j-1} p(x_j, y_k) \log_2 \left[\frac{1}{p(x_j \mid y_k)} \right]$$

Now, considering both the uncertainty conditions (before and after applying the inputs), we come to know that the difference, i.e. $H(x) - H(x \mid y)$ must represent the uncertainty about the channel input that is resolved by observing the channel output.

This is called as the Mutual Information of the channel.

Denoting the Mutual Information as $I(x; y)$, we can write the whole thing in an equation, as follows:

$$I(x; y) = H(x) - H(x \mid y)$$

Hence, this is the equation representation of Mutual Information.

Properties of Mutual Information

These are the properties of Mutual information:

- Mutual information of a channel is symmetric.

$$I(x;y) = I(y;x)$$

- Mutual information is non-negative.

$$I(x;y) \geq 0$$

- Mutual information can be expressed in terms of entropy of the channel output.

$$I(x;y) = H(y) - H(y|x)$$

Where, $H(y|x)$ is a conditional entropy.

- Mutual information of a channel is related to the joint entropy of the channel input and the channel output.

$$I(x;y) = H(x) + H(y) - H(x,y)$$

Where the joint entropy $H(x,y)$ is defined by,

$$H(x,y) = \sum_{j=0}^{j-1}\sum_{k=0}^{k-1} p(x_j, y_k) \log_2 \left(\frac{1}{p(x_i, y_k)} \right)$$

Channel Capacity

We have so far discussed mutual information. The maximum average mutual information, in an instant of a signaling interval, when transmitted by a discrete memory less channel, the probabilities of the rate of maximum reliable transmission of data, can be understood as the channel capacity.

It is denoted by C and is measured in bits per channel use.

Discrete Memoryless Source

A source from which the data is being emitted at successive intervals, which is independent of previous values, can be termed as discrete memoryless source.

This source is discrete as it is not considered for a continuous time interval, but at discrete time intervals. This source is memoryless as it is fresh at each instant of time, without considering the previous values.

Source Coding Theorem

The Code produced by a discrete memoryless source, has to be efficiently represented, which is an important problem in communications. For this to happen, there are code words, which represent these source codes.

For example, in telegraphy, we use Morse code, in which the alphabets are denoted by Marks and Spaces. If the letter E is considered, which is mostly used, it is denoted by "." Whereas the letter Q which is rarely used, is denoted by "--.-"

Let us take a look at the block diagram.

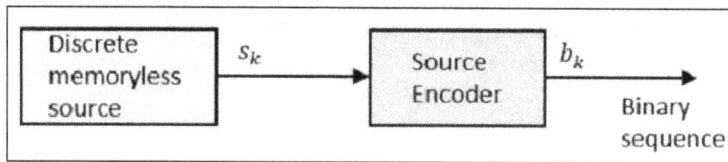

Where S_k is the output of the discrete memory less source and b_k is the output of the source encoder which is represented by 0s and 1s.

The encoded sequence is such that it is conveniently decoded at the receiver.

Let us assume that the source has an alphabet with k different symbols and that the kth symbol S_k occurs with the probability P_k, where k = 0, 1...k-1.

Let the binary code word assigned to symbol S_k, by the encoder having length l_k, measured in bits.

Hence, we define the average code word length $\bar{L}$ of the source encoder as:

$$\bar{L} = \sum_{k=0}^{k-1} P_k l_k$$

$\bar{L}$ represents the average number of bits per source symbol,

$$L_{min} = \text{minimum possible value of } \bar{L}$$

Then coding efficiency can be defined as,

$$\eta = \frac{L\,min}{\bar{L}}$$

With $\bar{L} \geq L_{min}$ we will have $\eta \leq 1$

However, the source encoder is considered efficient when $\eta = 1$

For this, the value L_{min} has to be determined.

Let us refer to the definition, "Given a discrete memoryless source of entropy $H(\delta)$, the average code-word length $\overline{L}$ for any source encoding is bounded as $\overline{L} \geq H(\delta)$."

In simpler words, the code word (example: Morse code for the word QUEUE is -.- ..- ...- .) is always greater than or equal to the source code (QUEUE in example). Which means, the symbols in the code word are greater than or equal to the alphabets in the source code.

Hence with $L_{min} = H(\delta)$, the efficiency of the source encoder in terms of Entropy $H(\delta)$ may be written as,

$$\eta = \frac{H(\delta)}{\overline{L}}$$

This source coding theorem is called as noiseless coding theorem as it establishes an error-free encoding. It is also called as Shannon's first theorem.

Channel Coding Theorem

The noise present in a channel creates unwanted errors between the input and the output sequences of a digital communication system. The error probability should be very low, nearly ≤ 10 for a reliable communication.

The channel coding in a communication system, introduces redundancy with a control, so as to improve the reliability of the system. The source coding reduces redundancy to improve the efficiency of the system.

Channel coding consists of two parts of action:

- Mapping incoming data sequence into a channel input sequence.
- Inverse Mapping the channel output sequence into an output data sequence.

The final target is that the overall effect of the channel noise should be minimized.

The mapping is done by the transmitter, with the help of an encoder, whereas the inverse mapping is done by the decoder in the receiver.

Channel Coding

Let us consider a discrete memoryless channel (δ) with Entropy H (δ).

T_s indicates the symbols that δ gives per second.

Channel capacity is indicated by C.

Channel can be used for every T_c secs.

Hence, the maximum capability of the channel is C/T_c

The data sent $= \dfrac{H(\delta)}{T_s}$.

If $\dfrac{H(\delta)}{T_s} \leq \dfrac{C}{T_c}$ it means the transmission is good and can be reproduced with a small probability of error.

In this, $\dfrac{C}{T_c}$ is the critical rate of channel capacity.

If $\dfrac{H(\delta)}{T_s} = \dfrac{C}{T_c}$ then the system is said to be signaling at a critical rate.

Conversely, if $\dfrac{H(\delta)}{T_s} = \dfrac{C}{T_c}$, then the transmission is not possible.

Hence, the maximum rate of the transmission is equal to the critical rate of the channel capacity, for reliable error-free messages, which can take place, over a discrete memoryless channel. This is called as Channel coding theorem.

Error Control Coding

Noise or Error is the main problem in the signal, which disturbs the reliability of the communication system. Error control coding is the coding procedure done to control the occurrences of errors. These techniques help in Error Detection and Error Correction.

There are many different error correcting codes depending upon the mathematical principles applied to them. But, historically, these codes have been classified into Linear block codes and Convolution codes.

Linear Block Codes

In the linear block codes, the parity bits and message bits have a linear combination, which means that the resultant code word is the linear combination of any two code words.

Let us consider some blocks of data, which contains k bits in each block. These bits are mapped with the blocks which has n bits in each block. Here n is greater than k. The transmitter adds redundant bits which are (n-k) bits. The ratio k/n is the code rate. It is denoted by r and the value of r is r < 1.

The (n-k) bits added here, are parity bits. Parity bits help in error detection and error correction, and also in locating the data. In the data being transmitted, the left most bits of the code word correspond to the message bits, and the right most bits of the code word correspond to the parity bits.

Systematic Code

Any linear block code can be a systematic code, until it is altered. Hence, an unaltered block code is called as a systematic code.

Following is the representation of the structure of code word, according to their allocation.

If the message is not altered, then it is called as systematic code. It means, the encryption of the data should not change the data.

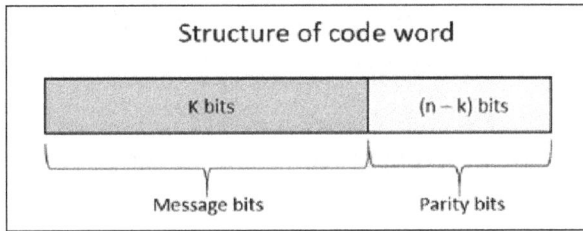

Structure of code word

K bits	(n − k) bits
Message bits	Parity bits

Convolution Codes

So far, in the linear codes, we have discussed that systematic unaltered code is preferred. Here, the data of total n bits if transmitted, k bits are message bits and (n-k) bits are parity bits.

In the process of encoding, the parity bits are subtracted from the whole data and the message bits are encoded. Now, the parity bits are again added and the whole data is again encoded.

The following figure quotes an example for blocks of data and stream of data, used for transmission of information.

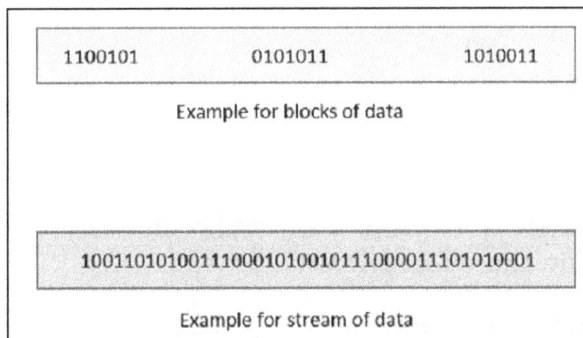

1100101   0101011   1010011

Example for blocks of data

1001101010011100010100101110000111101010001

Example for stream of data

The whole process, stated above is tedious which has drawbacks. The allotment of buffer is a main problem here, when the system is busy.

This drawback is cleared in convolution codes. Where the whole stream of data is assigned symbols and then transmitted. As the data is a stream of bits, there is no need of buffer for storage.

Hamming Codes

The linearity property of the code word is that the sum of two code words is also a code word. Hamming codes are the type of linear error correcting codes, which can detect up to two bit errors or they can correct one bit errors without the detection of uncorrected errors.

While using the hamming codes, extra parity bits are used to identify a single bit error. To get from one-bit pattern to the other, few bits are to be changed in the data. Such number of bits can be termed as Hamming distance. If the parity has a distance of 2, one-bit flip can be detected. But this can't be corrected. Also, any two bit flips cannot be detected.

However, Hamming code is a better procedure than the previously discussed ones in error detection and correction.

BCH Codes

BCH codes are named after the inventors Bose, Chaudari and Hocquenghem. During the BCH code design, there is control on the number of symbols to be corrected and hence multiple bit correction is possible. BCH codes is a powerful technique in error correcting codes.

For any positive integers $m \geq 3$ and $t < 2^{m-1}$ there exists a BCH binary code. Following are the parameters of such code.

      Block length $n = 2^m - 1$

      Number of parity-check digits $n - k \leq mt$

      Minimum distance $d_{min} \geq 2t + 1$

This code can be called as t-error-correcting BCH code.

Cyclic Codes

The cyclic property of code words is that any cyclic-shift of a code word is also a code word. Cyclic codes follow this cyclic property.

For a linear code C, if every code word i.e., $C = (C_1, C_2, \ldots\ldots C_n)$ from C has a cyclic right shift of components, it becomes a code word. This shift of right is equal to n-1 cyclic left shifts. Hence, it is invariant under any shift. So, the linear code C, as it is invariant under any shift, can be called as a cyclic code.

Cyclic codes are used for error correction. They are mainly used to correct double errors and burst errors.

Hence, these are a few error correcting codes, which are to be detected at the receiver. These codes prevent the errors from getting introduced and disturb the communication. They also prevent the signal from getting tapped by unwanted receivers. There is a class of signaling techniques to achieve this.

Spread Spectrum Modulation

A collective class of signaling techniques are employed before transmitting a signal to provide a secure communication, known as the Spread Spectrum Modulation. The main advantage of spread spectrum communication technique is to prevent "interference" whether it is intentional or unintentional.

The signals modulated with these techniques are hard to interfere and cannot be jammed. An intruder with no official access is never allowed to crack them. Hence, these techniques are used for military purposes. These spread spectrum signals transmit at low power density and has a wide spread of signals.

Pseudo-noise Sequence

A coded sequence of 1s and 0s with certain auto-correlation properties, called as Pseudo-Noise coding sequence is used in spread spectrum techniques. It is a maximum-length sequence, which is a type of cyclic code.

Narrow-band and Spread-spectrum Signals

Both the Narrow band and Spread spectrum signals can be understood easily by observing their frequency spectrum as shown in the following figures.

Narrow-band Signals

The Narrow-band signals have the signal strength concentrated as shown in the following frequency spectrum figure.

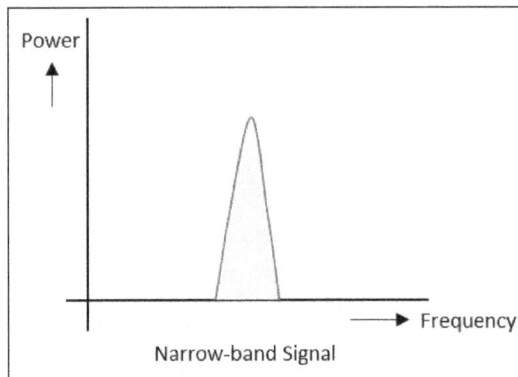

Narrow-band Signal

Following are some of its features:

- Band of signals occupy a narrow range of frequencies.
- Power density is high.
- Spread of energy is low and concentrated.

Though the features are good, these signals are prone to interference.

Spread Spectrum Signals

The spread spectrum signals have the signal strength distributed as shown in the following frequency spectrum figure.

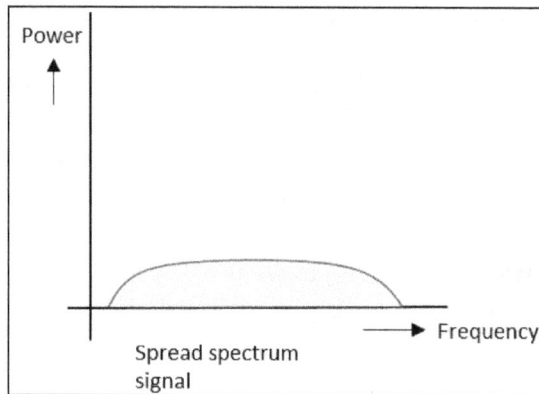

Following are some of its features:

- Band of signals occupy a wide range of frequencies.
- Power density is very low.
- Energy is wide spread.

With these features, the spread spectrum signals are highly resistant to interference or jamming. Since multiple users can share the same spread spectrum bandwidth without interfering with one another, these can be called as multiple access techniques.

FHSS and DSSS/CDMA

Spread spectrum multiple access techniques uses signals which have a transmission bandwidth of a magnitude greater than the minimum required RF bandwidth.

These are of two types.

- Frequency Hopped Spread Spectrum (FHSS).
- Direct Sequence Spread Spectrum (DSSS).

Frequency Hopped Spread Spectrum

This is frequency hopping technique, where the users are made to change the frequencies of usage, from one to another in a specified time interval, hence called as frequency hopping. For example, a frequency was allotted to sender 1 for a particular period of time. Now, after a while, sender 1 hops to the other frequency and sender 2 uses the first frequency, which was previously used by sender 1. This is called as frequency reuse.

The frequencies of the data are hopped from one to another in order to provide a secure transmission. The amount of time spent on each frequency hop is called as Dwell time.

Direct Sequence Spread Spectrum

Whenever a user wants to send data using this DSSS technique, each and every bit of the user data is multiplied by a secret code, called as chipping code. This chipping code is nothing but the spreading code which is multiplied with the original message and transmitted. The receiver uses the same code to retrieve the original message.

Comparison between FHSS and DSSS/CDMA

Both the spread spectrum techniques are popular for their characteristics. To have a clear understanding, let us take a look at their comparisons.

FHSS	DSSS / CDMA
Multiple frequencies are used.	Single frequency is used.
Hard to find the user's frequency at any instant of time.	User frequency, once allotted is always the same.
Frequency reuse is allowed.	Frequency reuse is not allowed.
Sender need not wait.	Sender has to wait if the spectrum is busy.
Power strength of the signal is high.	Power strength of the signal is low.
Stronger and penetrates through the obstacles.	It is weaker compared to FHSS.
It is never affected by interference.	It can be affected by interference.
It is cheaper.	It is expensive.
This is the commonly used technique.	This technique is not frequently used.

Advantages of Spread Spectrum

Following are the advantages of spread spectrum:

- Cross-talk elimination,

- Better output with data integrity,

- Reduced effect of multipath fading,

- Better security,

- Reduction in noise,

- Co-existence with other systems,

- Longer operative distances,

- Hard to detect,

- Not easy to demodulate/decode,

- Difficult to jam the signals.

Although spread spectrum techniques were originally designed for military uses, they are now being used widely for commercial purpose.

Baseband Systems

Baseband transmission is the simplest form for the communication of information. Discrete information is communicated with specific symbols selected from a finite set of symbols. In baseband transmission, symbols are simply communicated as a pulse with a discrete voltage level and, for binary transmission, only two voltages are used. A series of pulses forms a pulse train that carries the full message. Prior to transmission, especially in radio systems, these pulses are shaped to limit their high frequency content so as to minimize crosstalk with adjacent communication channels. During transmission through a band limited channel, pulses are dispersed (spread) in time and can overlap with each other giving rise to intersymbol interference (ISI). When pulses reach the receiver, dispersion and other distortions can be partially compensated with an equalizer.

Impulse Response in a Bandlimited Channel

We first consider a series of narrow symbol pulses. Restricted channel bandwidth, disperses (or spreads) the pulse in time and necessitates an increased interval between symbols. The maximum rate at which symbols can be sent is proportional to channel bandwidth. Pulse dispersion relates directly to channel impulse response which can be determined through the Fourier transform of H(f), the channel frequency response.

We now consider a single rectangular pulse with amplitude h and interval τ. The amplitude spectrum of the pulse is determined using the Fourier transform. As illustrated below, the spectrum has magnitude hτ at zero frequency and a sin $\pi\tau f$ /$\pi\tau f$ variation with frequency. In addition to amplitude spectrum, there is a phase spectrum (not shown) that has values 0 and π when the pulse is centered on the t = 0 axis. Note the spectral nulls at $f = 1/\tau, 2/\tau, 3/\tau, \ldots$.

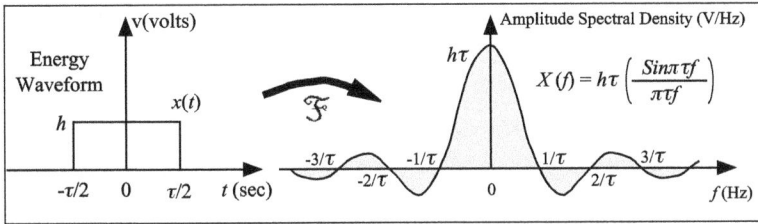

Fourier transform of a single pulse waveform.

Note that if the weight (i.e. area) of the pulse, $h\tau$, is held constant while the width is decreased, the spectral peak remains constant at $h\tau$ and the spectrum spreads out in frequency. Taken to the limit, the pulse approaches an impulse and the spectrum becomes "flat".

Next we consider a narrow rectangular pulse passing through a bandlimited transmission system modeled by an ideal lowpass filter (LPF). If the input pulse spectrum is approximately "flat", and the transmission system has a rectangular "brickwall" frequency response, the receiver will see a rectangular spectrum centered about 0 Hz. Applying the inverse Fourier transform to this rectangular spectrum, we determine that the received voltage response has peak amplitude $2fch\tau$ and $\sin \pi fct/\pi fct$ variation with time.

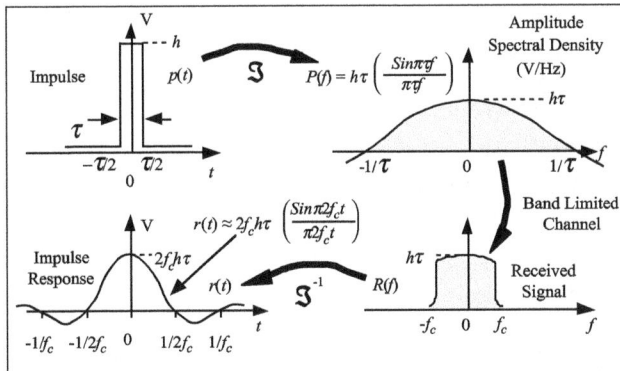

Impulse response of a bandlimited channel.

Note that the impulse response tails oscillate at the LPF cutoff frequency. This can be related to a practical low pass filter where transfer function poles nearest the cutoff frequency have the lowest damping ratio (i.e. high Q) and, following an impulse, the filter continues to oscillate at the cutoff frequency because of the high Q poles.

The integrals $\int_0^\infty \dfrac{\sin x}{x}dx = \dfrac{\pi}{2}$ and $\int_0^\infty \left(\dfrac{\sin x}{x}\right)^2 dx = \dfrac{\pi}{2}$ are useful for analysis of the above pulse spectra and impulse response.

Maximum Signaling Rate

Message symbols occur in a sequence at rate R_s called the baud rate of the transmission; one band equals one symbol per second. With limited channel bandwidth (as in a low-pass filter), symbol pulses are spread out in time (dispersed) and, if interference between successive pulses is to be avoided, there must be a minimum interval between pulses. The maximum symbol rate, R_{max}, is therefore limited by the channel bandwidth. We shall see that by setting the symbol rate equal to twice the lowpass filter bandwidth and by sampling the received signal at appropriate times, we have the maximum symbol rate that can be attained without intersymbol interference.

Impulse Signaling with an Ideal Filter

Assume a first impulse (not a rectangular pulse) transmitted through an ideal lowpass channel with unity gain from 0 Hz to the cutoff frequency f_c. The received signal (the channel impulse response) will be of the form $\sin f_c t / f_c t$ with a main lobe having peak voltage at time t_1 and with zero crossings at time intervals $1/2f_c$ about time t_1. If the first impulse is followed by a second impulse, transmitted after delay $1/2f_c$, the second impulse response will have a main lobe with voltage peak at time t_2 occurring at a zero crossing of the first impulse response. The second symbol may then be received without interference provided that the received signal is sampled precisely at the zero crossing of the first impulse response. Further zero crossings of both impulse responses will be coincident at intervals of $1/2f_c$ and, if additional impulse symbols are transmitted at time intervals of $1/2f_c$, these too can be received without ISI.

Symbols may be communicated without inter-symbol interference at a rate of $f_{sy} = 2f_c$. This symbol rate, f_{sy}, is known as the Nyquist signaling rate.

In figure, five impulses with weights +3 and +1 are shown after transmission through a lowpass channel with cutoff frequency at 500 kHz. The received voltage is the superposition of all impulse responses from the transmitted symbol sequence. A sample taken at time t = 8 us, for example, contains no output from pulses p_1, p_2 p_4 and p_5 and is therefore responsive only to the amplitude of p_3. This condition of zero ISI is independent of the other pulse amplitudes.

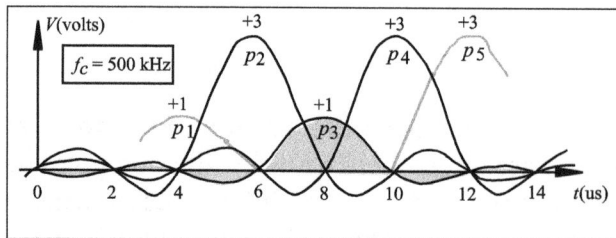

Intersymbol Interference.

For zero ISI we must exactly match the channel bandwidth to the reciprocal of the signaling rate. In this case, the channel includes the transmit filter, the transmission link and the receive filter. In practice, we use an interconnecting transmission link with somewhat larger bandwidth and then control the channel bandwidth with accurate transmitter and receiver filters with cutoff at fc. Precise cut-off frequencies can be implemented with clocked digital filters.

Impulse Signaling with Practical Filters

The sharp cut-off, rectangular "brickwall" filters illustrated in the previous section cannot be implemented in practice; the transition from passband to stopband must occur over some frequency range. Filters are implemented with a roll-off that is symmetric about f_c extending up to $(1+r)f_c$ where r is the channel roll-off factor. The transition region characteristic usually approximates the first 180° of a raised cosine leading to the moniker "raised cosine filter". With roll-off factor r ≈ 0.3, the transmitted spectrum is 30% in excess of what would be transmitted with a "brickwall" filter (i.e. 30% "excess bandwidth").

The gradual filter transition results shortening the ripple "tails" in the channel impulse response. This reduction of tail amplitude and duration significantly reduces ISI and relaxes the need for precise matching of signaling rate to the zero crossing rate. For small timing offsets, the previous impulse has a zero crossing near the desired sampling time. On the other hand, interfering impulses from the distant past may have displaced zero crossings however their amplitude has now become insignificant.

Figure illustrates excess bandwidth and reduced impulse response duration. The filter characteristic follows a raised cosine function in the transition region and has a gain of 0.5 (-6dB) at the frequency f_c. The total transmission bandwidth required is $f_b = (1+r)f_c$.

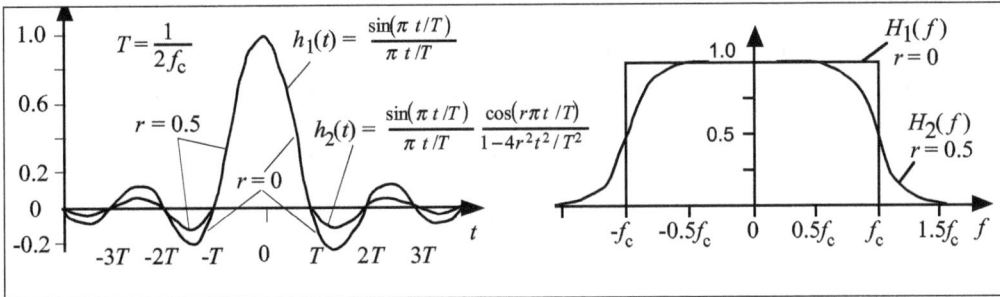

Raised cosine frequency response and impulse response (r = 0.5).

It is not possible to create a filter with perfectly sharp cut off in the frequency domain - all practical filters must have some excess bandwidth. The raised cosine pulse is defined in frequency by:

$$H_2(f) = 1 \qquad\qquad 0 \le |f| \le \frac{1-r}{2T}$$

$$H_2(f) = \cos\left[\frac{\pi T}{2\pi}\left(f - \frac{1-r}{2T}\right)\right] \qquad\qquad \frac{1-r}{2T} < |f| \le \frac{1+r}{2T}$$

$$H_2(f) = 0 \qquad\qquad \frac{1+r}{2T} \le |f| \le \infty$$

Additionally, an ideal filter must have an impulse response that extends to infinite time before and after the pulse peak and, if the filter is to be causal (output response occurs after the input is applied), the peak output would occur an infinite time after the input. This would certainly not make for a useful communication system since one would need to wait more than a lifetime for the message to reach the receiver.

To this point we have considered filter structures (transversal filters, for example) with constant delay and therefore phase shift that increases linearly with frequency. Recursive digital filters and most analog filters do not have constant delay throughout the passband so the output impulse response is not symmetrical about the peak value. In rough terms, the ripples preceding the main lobe are eliminated and the ripples following the main lobe are approximately doubled in an analog filter impulse response.

Signaling with Rectangular Pulses

To this point we have considered the transmission of impulses (extremely short duration and extremely high amplitude). Many digital transmission systems transmit finite amplitude pulses of duration equaling the symbol interval or half the symbol interval. The pulse response presented to the receiver can be estimated by three methods:

- Convolve the impulse response with the transmitted pulse,

- Divide the transmitted pulse into narrow pulses determine the approximate impulse response of each then add (superposition) to get the received pulse response, and

- Differentiate the transmitted pulse to obtain leading and trailing impulses, obtain their impulse responses the integrate to get the pulse response.

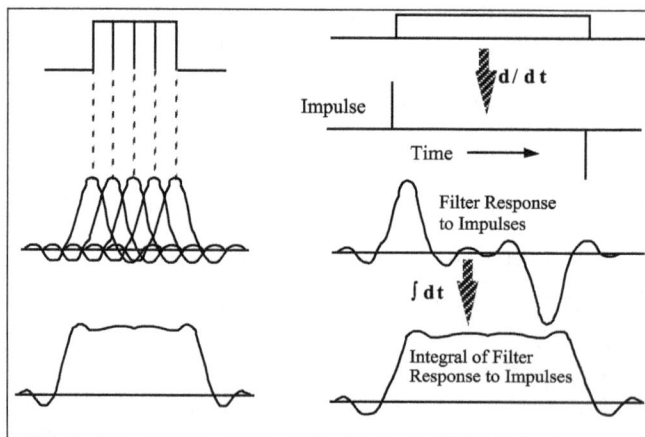

Calculation of pulse response.

Another method of analyzing rectangular transmitted pulses is to consider the process of flat-top sampling and to recall the inverse $\sin \pi fT / \pi fT$ filter used to reverse the effects of flat-top reconstruction. With a similar inverse filter in the pulse transmission path, the received pulse response can be made to have the same waveshape (and zero crossings) as the channel impulse response.

Although wireline transmission systems such as Ethernet and digital telephone circuits transmit rectangular pulses directly, more elegant techniques are required for advanced communication systems. Pulses are shaped prior to transmission and this shaping becomes part of the channel frequency response.

Signaling with Root-raised-cosine Pulses

By allowing a little extra bandwidth in the channel's equivalent lowpass filter, many difficulties associated with the "sinc" channel impulse response have been mitigated. In a communication system, there is need to associate some of this channel equivalent lowpass filter with the transmitter and some with the receiver. Transmit filtering is need to limit high frequency components in the symbol pulses and avoid interference with other communication channels. At the receiver, filtering is needed to remove incoming out-of-band noise prior to the detection process. In several cases, especially in wireless transmission, the physical channel is essentially "flat" and all of the channel filtering occurs in the transmitter and receiver. Since the transfer function of the overall channel is the product of the cascaded transmit filter and receive filter, each filter is implemented with a "root-raised-cosine" characteristic.

To this point, we assume that the transmitter generates an impulse then filters it prior to transmission. The transmitter is substantially simplified by directly generating the root-raisedcosine transmitted pulse. A reasonable truncated approximation of the desired pulse is synthesized from a series of stored samples covering the range 4-6 cycles on either side of the main lobe. A time domain expression for the root-raised-cosine pulse is,

$$v(t) = \frac{1}{\sqrt{T}} \frac{\sin[\pi(1-r)t/T + (4rt/T)\cos[\pi(1+r)t/T]]}{(\pi t T)[1-(4rt/T)^2]} \qquad t \neq 0, t \neq \pm\frac{T}{4r}$$

$$v(t) = \frac{1}{\sqrt{2T}}\left(1 - r + \frac{4r}{\pi}\right) \qquad t = 0$$

$$v(t) = \frac{r}{\sqrt{2T}}\left[\left(1+\frac{2}{\pi}\right)\sin\left(\frac{\pi}{4r}\right)+\left(1-\frac{2}{\pi}\right)\cos\left(\frac{\pi}{4r}\right)\right] \qquad t = \pm\frac{T}{4r}$$

Pulse Spectrum and Eye Patterns

We extend our study of pulse spectral density to a series of random pulses as would be used for message transmission. This spectrum is truncated by channel filters that are used to smooth the transmitted waveform and to eliminate much of the noise that would enter the receiver. The effect of channel filters on the received signal waveform is studied with the aid of a display known as an eye pattern. We begin by a calculation of the spectrum of a rectangular binary pulse sequence.

Spectrum of a Continuous Pulse Sequence

Assume a single rectangular pulse, $w(t)$, of amplitude A and period T centered about $t = 0$. The amplitude spectral density $W(f)$ is calculated using the Fourier Transform and the normalized energy spectral density $E(f)$ is simply the square of the Fourier Transform.

$$W(f) = AT\left(\frac{\sin \pi Tf}{\pi Tf}\right), \; E_w(f) = A^2 T^2 \left(\frac{\sin \pi Tf}{\pi Tf}\right)^2$$

Since power is energy per unit time, the single pulse power spectral density, S(f), is,

$$S_w(f) = \frac{E(f)}{T} = A^2 T \left(\frac{\sin \pi Tf}{\pi Tf}\right)^2$$

In a sequence of pulses with random amplitude, spectral components of the message will add on a power basis. Power spectral density $S_m(f)$ is the time average of the PSDs for each pulse.

$$PSD = (S_w(f) >= T\left(\frac{\sin \pi Tf}{\pi Tf}\right)^2 < A_n^2 > n = 1, 2, 3...$$

$$S_m(f) = A^2 T \left(\frac{\sin \pi Tf}{\pi Tf}\right)^2 \quad \text{if} \quad A_n = A, -A$$

When the pulse sequence is periodic, the PSD will be a line spectrum rather than a continuous spectrum. A square alternating sequence is a simple example.

Eye Pattern Measurement

The effect of bandwidth limitation, dispersion, distortion, intersymbol interference and timing impairments can be studied with the aid of an eye pattern display. This display is produced on an oscilloscope by superimposing many traces of the received signal and, when binay data is used, the resulting pattern resembles a human eye. At the sampling instant, the received signal should be well above or below the threshold voltage to ensure reliable detection of a binary one or zero and thus the eye opening indicates margin against possible errors caused by noise. It is evident that the best position for the receiver voltage threshold and sampling instant is in the center of the eye opening.

The display shows one central symbol plus part of the preceding symbol and part of the following symbol. Since the oscilloscope is triggered by the data generator clock, the superposition of all possible data sequences can be shown on the screen. The eye diagram thus displays the 8 possible trajectories for 3 bits. It is preferable to operate the oscilloscope in storage mode.

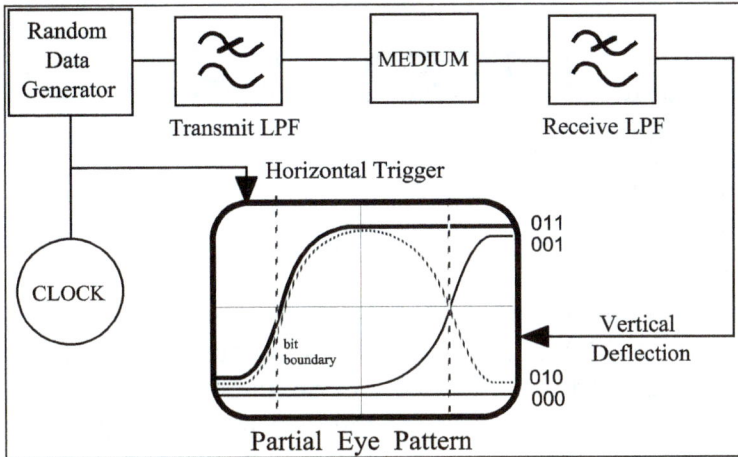

Eye pattern oscilloscope display.

Elimination of Higher Spectral Frequencies

Channel lowpass filters eliminate the higher frequency components of the transmitted pulse sequence and smooth the sharp symbol transitions initially assumed. As the filter cut-off frequency approaches one half the symbol rate, the rise times become longer, the pulse broadens and the impulse response extends into the time allocated for the next symbol. This gives rise to the well-known eye diagram, illustrated below in binary form.

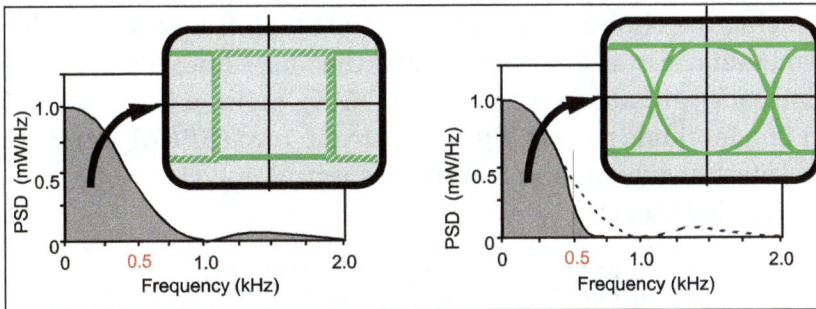

Eye diagram after elimination of higher spectral frequencies.

Loss of Low Frequencies

Many transmission systems introduce a component, such as a transformer, that eliminates a small portion of the spectral density extending upwards form zero hertz. This small loss of spectral power seems at first to be innocuous, however, it has a major effect on the eye pattern of an uncoded binary transmission. The beginning point of each trajectory in the eye diagram is influenced by the preamble of bits leading up to the 3 bit times shown in the display. In systems where there is loss of transmission at low frequencies, a positive preamble will result in a negative displacement of the trajectory while a negative preamble will have the opposite effect. This variation in the trajectories reduces the eye opening.

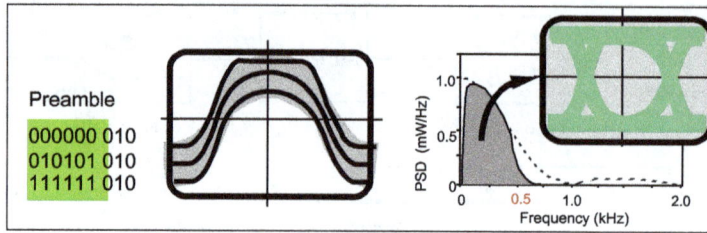

Eye diagram after loss of low frequencies.

Transmission systems are tested with pseudo-random sequences. (These are also known as pseudo-noise (PN) data sequences). Commercial test units provide a variety of PN sequence lengths and one choice might be $2^{11}-1 = 2047$. Most instruments can provide much longer sequences. Although the $2^{11}-1$ sequence repeats every 2047 bits, the longest string of sequential ones (or zeros) is 11. The PN code provides the complete variety of 11-bit preambles.

Baseband Line Coding

Binary digital signals may be formatted such that they are more suitable for transmission on a particular medium such as a transmission line. This formatting is referred to as line coding and several examples are illustrated in figure. Line coding may simply define the pulse shape for an individual symbol (for example NRZ or RZ) or it may also define the pulse sequence format for successive binary symbols (for example BPRZ). Factors influencing the choice of line code include high frequency and low frequency cut-off of the medium, signal-tonoise ratio and the phase linearity of the channel. Popular 2 level line codes are non-return to zero (NRZ), biphase (Bi∅) and delay modulation. A common 3 level line code is bipolar return to zero (BPRZ), a code that incorporates alternate-mark-inversion and features a spectral null at zero hertz. This make is suitable for ac coupled media.

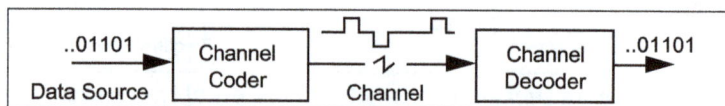

Baseband Line Coder/Decoder.

A selection of line codes can be generated by the following simple logic circuits. Conventional silicon logic circuits generate a unipolar output (for example, 0v and 5v) that must be level translated to give a polar line signal (+2.5v and -2.5v).

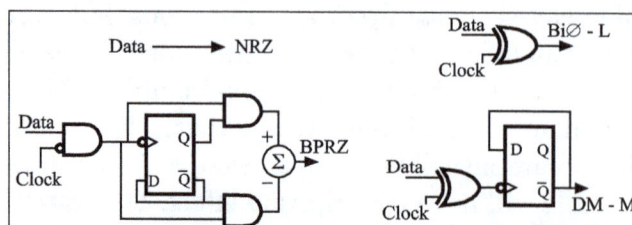

Baseband line encoder circuits.

Frequency Spectrum of Line Codes

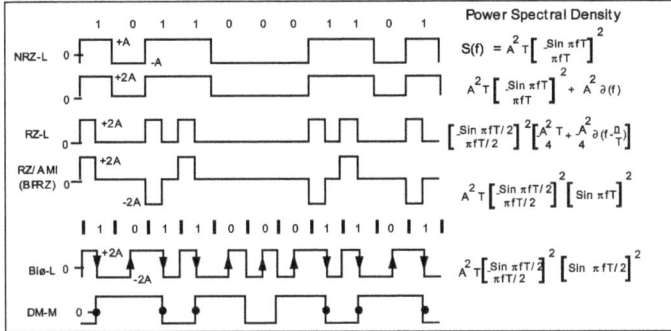

Line Code Waveforms.

Waveforms for common line codes are shown below along with an expression for the power spectral density, $S(f)$, for positive and negative frequencies. The pulse symbol rate is $1/T$ and 50% pulse width ($T/2$) is assumed for RZ and BPRZ. Note that A is not always equal to the peak pulse voltage.

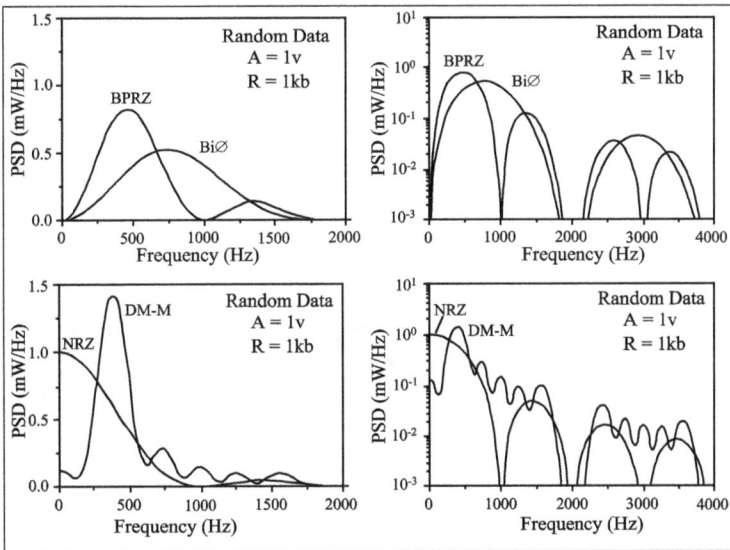

Line Code Power Spectral Density (PSD).

Application of the Biphase and BPRZ Codes

The biphase (Biø) line code is also widely known as the Manchester code. It has widespread application in 10 Mb/s Ethernet transmission both on coaxial cable and on twisted pair. The very low spectral content near zero hertz and the high bit rate allow transmission lines to be coupled to the electronic circuits through very small (and cheap) transformers. Another feature of this code is the regular voltage transitions that facilitate recovery of date rate clock in the receiver. The Local Talk transmission format, standard on Apple computers for many years, uses the (Biø) line code together with differential encoding.

The bipolar-return-to-zero (BPRZ) line code has been widely applied for paired wire transmission of 1.544 Mb/s DS1 signals in the telephone system. These signals are transmitted on wire sections as long as 1.8 km (6000 feet) and there are potentially damaging foreign voltages introduced in the transmission link. Transformer coupling is used to protect transmitter and receiver electronics from lighting induced surges and ground potential differences arising from 60 Hz power distribution. Although both codes have a spectral null at zero hertz, BPRZ was chosen over BiØ because, for the same bit rate, BPRZ requires less of the high frequency spectrum which is greatly attenuated in long transmission lines. Another advantage of dc free codes is that there is less spectral energy in the low frequency voice band. When compared to NRZ coding, this results in less interference to voice signals due to crosstalk between adjacent wire pairs.

BPRZ Eye Pattern with Loss of Low Frequencies

The following eye patterns show the two level NRZ and the three level BPRZ cases. For the BPRZ code and there is need for two thresholds and this results in less voltage margin for errors introduced by noise.

The robustness of the BPRZ line code to loss of low frequency transmission is shown in the series of patterns resulting from progressive removal of low frequency content. For the 255 bit PN code sequence used in these tests, BPRZ can withstand at least 10 times more bandwidth loss at low frequencies. For longer PN data sequences, the NRZ code has even more problems with removal of low frequencies.

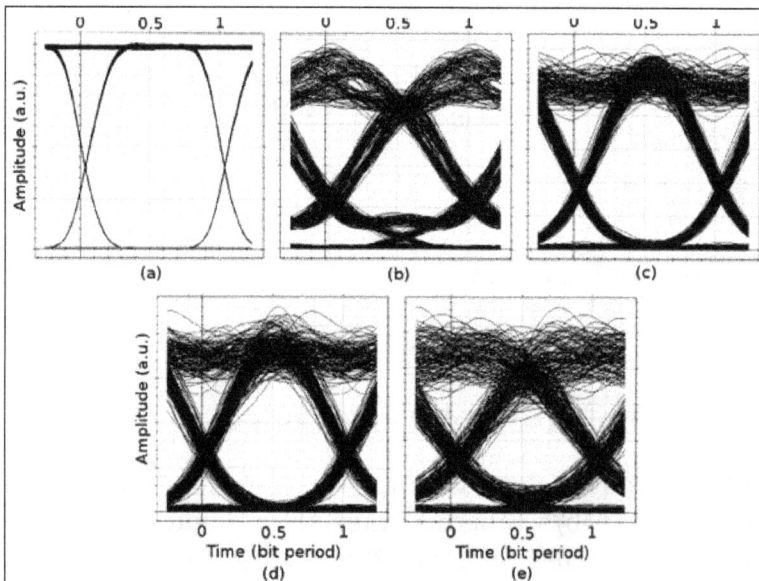

Loss of low frequencies in NRZ and BPRZ eye patterns.

Differential Coding

In twisted pair baseband systems and in BPSK carrier systems it is convenient if the data can be correctly decoded even when there is a polarity reversal in the transmission medium. Bipolar return to zero (BPRZ) coding is inherently able to function with reversed polarity. Other coding formats can be differentially coded. A disadvantage of this method is that one transmission error will cause two decoded errors.

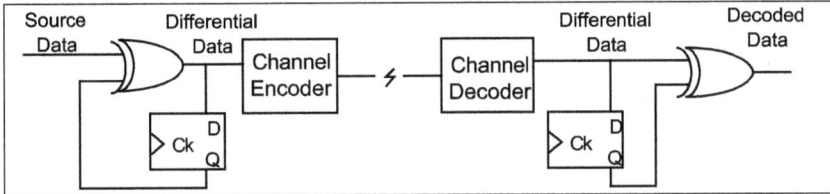

Differential encoder/decoder (Mark).

In differential "mark" encoding, a logic one in the source data results in a transition in the differential data. In "space" encoding, a logic zero in the source data results in a transition in the differential data.

PN Sequences, Scrambling and Error Measurement

Pseudorandom sequences are periodic signals that are deterministic and not al all random. Nevertheless, these signals appear to have the statistical properties of a random signal (such as sampled white noise). Pseudorandom binary sequences (PRBS) are used in testing to "exercise" a communication system by providing a variety of data sequences with a "full" frequency spectrum. Because the sequences are deterministic and fully predictable by the receiver, the error rate of the system can be measured while the PRBS is being sent. Another use of these sequences is to randomize message information making it more suitable for transmission.

Pseudorandom Sequence Generation

A PBRS sequence generator uses an n bit shift register with a feedback structure containing modulo-2 adders (i.e. exclusive OR gates) and connected to appropriate taps on the shift register. The generator generates a maximal length binary sequence of length 2n-1. The maximal length (or "m" sequence) has nearly random properties and is classed as a pseudo noise (PN) sequence. Properties of "m" sequences are as follows:

1. The Balance Property - In each period of the sequence, the number of '1's and the number of '0's differ by at most one. (In a 31 bit sequence, there are 16 '1's and 15 '0's).

2. The Run Property - Among the runs of '1's and of '0's in each period, one half the runs of each kind are of length one, one quarter are of length two, one eighth are of length three, etc. as long as these fractions give meaningful numbers of runs. (In a 31-bit sequence there are 16 runs).

3. Shift and add property - The modulo-2 sum of an "m" sequence and any cyclic shift of the same sequence results in a third cyclic shift of the same sequence.

4. The Correlation Property - If a full period of the sequence is compared, term-by-term, with any cyclic shift of itself, the number disagreements is one more than the number of agreements.

5. Spectral Properties - The sequence is periodic, and therefore the spectrum consists of a sequence of equally-spaced harmonics where the spacing is the reciprocal of the period. Prior to sinx/x shaping due to 100% pulse width, and with the exception of the dc harmonic, the magnitude of all the harmonics are equal. Aside from the spectral lines, the frequency spectrum of a maximal length sequence resembles that of a random sequence.

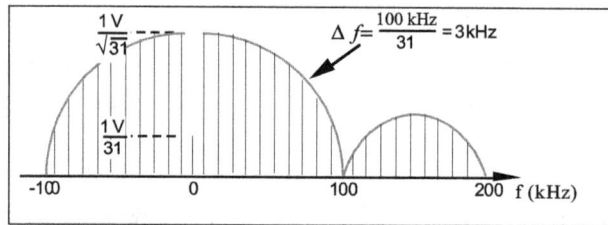

Example spectrum of a length 31 "m" sequence.

The shift register generator normally generates a maximal length sequence, however, it can also produce an alternate sequence of constant logic zero. With specific tap settings the maximal length (2^n-1) sequence is generated. Tap settings other than those indicated in the following table result in a shorter length sequence and the alternate sequence is correspondingly longer. These non-maximal length sequences do not have pseudorandom properties.

The table below lists required tap settings for a maximal length sequence. All possible "forward sequence" tap settings are listed for register lengths up to 8 and only selected coefficients are given for register lengths 9, 10 and 11. Reverse sequences can be generated by using "symmetric" taps. ie, $x^6 + x^5$ instead of $1 + x + x^6$.

Table: PN Generator Taps (Primitive Polynomials).

Register Length(n)				
2	[1, 2]			
3	[1,3]			
4	[1, 4]			
5	[2, 5]	[1, 2, 4, 5]	[2, 3, 4, 5]	
6	[1, 6]	[1, 2, 5, 6]	[2, 3, 5, 6]	
7	[3, 7]	[1, 2, 3, 7]	[1, 2, 4, 5, 6, 7]	[2, 3, 4, 7]
	[1, 2, 3, 4, 5, 7]	[2, 4, 6, 7]	[1, 7]	[1, 3, 6, 7]
	[2, 5, 6, 7]			

8	[2, 3, 4, 8]	[3, 5, 6, 8]	[1, 2, 5, 6, 7, 8]	
	[1, 3, 5, 8]	[2, 5, 6, 8]	[1, 5, 6, 8]	
	[1, 2, 3, 4, 6, 8]	[1, 6, 7, 8]		
9	[4, 9]	[3, 4, 6, 9]	[4, 5, 8, 9]	
	[1, 4, 8, 9]	[2, 3, 5, 9]	[1, 2, 4, 5, 6, 9]	
	[5, 6, 8, 9]	[1, 3, 4, 6, 7, 9]	[2, 7, 8, 9]	
10	[3, 10]	[2, 3, 8, 10]	[3, 4, 5, 6, 7, 8, 9, 10]	
	[1, 2, 3, 5, 6, 10]	[2, 3, 6, 8, 9, 10]	[1, 3, 4, 5, 6, 7, 8, 10]	
11	[2, 11]	[2, 5, 8, 11]	[2, 3, 7, 11]	

Scrambling

Scrambling is a randomizing method that achieves dc balance and breaks up long sequences of zero's or ones that might originate from a message source such as a computer keyboard. Long strings of zeros or ones introduce baseline wander in the receiver eye pattern and present difficulty for timing recovery systems that require frequent transitions. In scramblers, a maximal length sequence is modulo-2 added to the data sequence and this is attractive because it requires no extra bandwidth. Since the PN sequence is modulo-2 added to the message sequence, scrambling is the equivalent of modulation by the set of harmonics and we expect the result to be approximately white. This "whitening" disperses any large low frequency signal component that would seriously degrade transmission in a system with ac coupling. A serious problem can occur if the input stream has period equal the PN sequence since the scrambled stream itself can then have a large d.c. component resulting in severe baseline wander at the receiver. If the PN sequence has large period, this pathological situation will arise with very low probability.

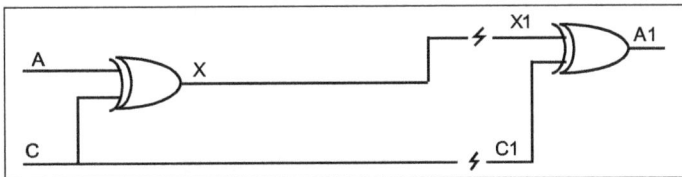

Polarity reversal using X-OR gate.

There are two forms of scrambling - self-synchronizing and frame-synchronized. Both types of scramblers use maximal-length shift-register sequences and in this discussion we will focus on the self-synchronized scrambler. The input data stream is modulo-2 added in the feedback path of a PN sequence generator as illustrated in Figure. At the receiver descrambler, the received stream is applied to the input of an identical shift-register. Since both shift-registers have identical inputs (in the absence of transmission errors), their outputs (C and C1) will be identical and, from the operation of the modulo-2 adders, it follows that the output stream will be identical to the input data stream.

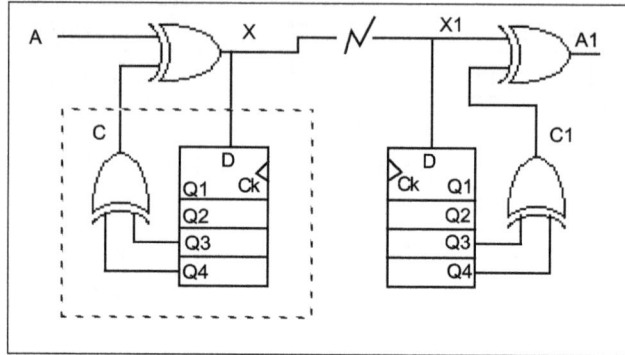

Self-synchronized scrambler/descrambler.

Two problems exist with this simple scrambling structure. The first problem is that a transmission error causes an initial error in the output A1 and then, after traveling through the shift register, it causes two more errors in the output. In this case there is error multiplication by a factor of three. The second problem has to do with scrambler lockup. When the input A is logic zero, the scrambler generates a PN sequence and the shift register takes on all states except the all zero state. When the input A is logic one, the register takes on all states except the all one's state. During this latter case, an input transition can occur while the register has all zero's and after this input transition, the scrambler remains in the all zero state (locked up) until there is another input transition. The output descrambler works perfectly under these conditions, however, the transmitted data is no longer randomized and the receiver may make errors due to baseline wander and the loss of clocking transitions.

Error Measurement Test Set

An elegant application of the scrambler concept is used in the measurement of errors on a digital transmission link. The data input to the transmitter scrambler is held at 'o' thus the descrambler output should always be 'o' in the absence of channel errors. When the switch in figure is in position 'a' the circuit operates as a self-synchronized descrambler. After some period of operation without errors, the receiver shift register is known to have an error free sequence and can therefore re-circulate independently by moving the switch to position "b". In this latter mode, the counter registers the true number of errors since the errored sequence no longer enters the shift register.

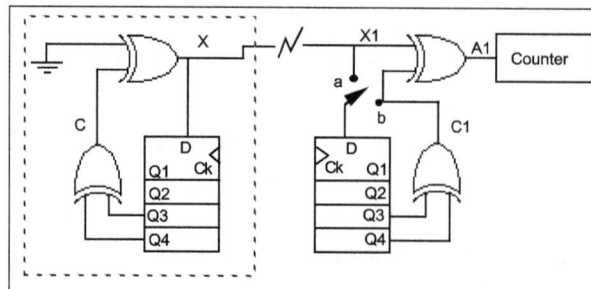

BER test setup.

Sampling Process

Sampling is defined as, "The process of measuring the instantaneous values of continuous-time signal in a discrete form". Sample is a piece of data taken from the whole data which is continuous in the time domain.

When a source generates an analog signal and if that has to be digitized, having 1sand 0s i.e., High or Low, the signal has to be discretized in time. This discretization of analog signal is called as Sampling.

The following figure indicates a continuous-time signal x (t) and a sampled signal x_s(t). When x (t) is multiplied by a periodic impulse train, the sampled signal x_s (t) is obtained.

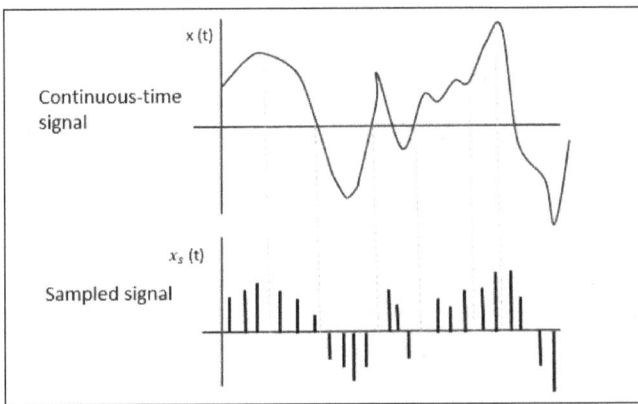

Sampling Rate

To discretize the signals, the gap between the samples should be fixed. That gap can be termed as a sampling period T_s.

$$Sampling\ Frequency = \frac{1}{T_s} = f_s$$

Where,

- T_s is the sampling time.

- f_s is the sampling frequency or the sampling rate.

Sampling frequency is the reciprocal of the sampling period. This sampling frequency, can be simply called as Sampling rate. The sampling rate denotes the number of samples taken per second, or for a finite set of values.

For an analog signal to be reconstructed from the digitized signal, the sampling rate should be highly considered. The rate of sampling should be such that the data in the message signal should neither be lost nor it should get over-lapped. Hence, a rate was fixed for this, called as Nyquist rate.

Nyquist Rate

Suppose that a signal is band-limited with no frequency components higher than W Hertz. That means, W is the highest frequency. For such a signal, for effective reproduction of the original signal, the sampling rate should be twice the highest frequency.

Which means,

$$fs = 2W$$

Where,

- fs is the sampling rate.
- W is the highest frequency.

This rate of sampling is called as Nyquist rate.

A theorem called, Sampling Theorem, was stated on the theory of this Nyquist rate.

Sampling Theorem

The sampling theorem, which is also called as Nyquist theorem, delivers the theory of sufficient sample rate in terms of bandwidth for the class of functions that are band-limited. The sampling theorem states that, "a signal can be exactly reproduced if it is sampled at the rate f_s which is greater than twice the maximum frequency W".

To understand this sampling theorem, let us consider a band-limited signal, i.e., a signal whose value is non-zero between some −W and W Hertz.

Such a signal is represented as $x(f) = 0$ for $|f| > W.$

For the continuous-time signal x (t), the band-limited signal in frequency domain, can be represented as shown in the following figure.

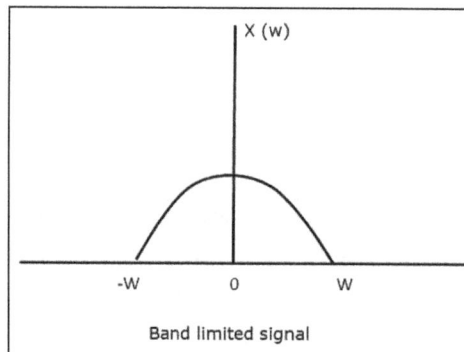

Band limited signal

We need a sampling frequency, a frequency at which there should be no loss of information, even after sampling. For this, we have the Nyquist rate that the sampling frequency should be two times the maximum frequency. It is the critical rate of sampling.

If the signal x(t) is sampled above the Nyquist rate, the original signal can be recovered, and if it is sampled below the Nyquist rate, the signal cannot be recovered.

The following figure explains a signal, if sampled at a higher rate than 2w in the frequency domain.

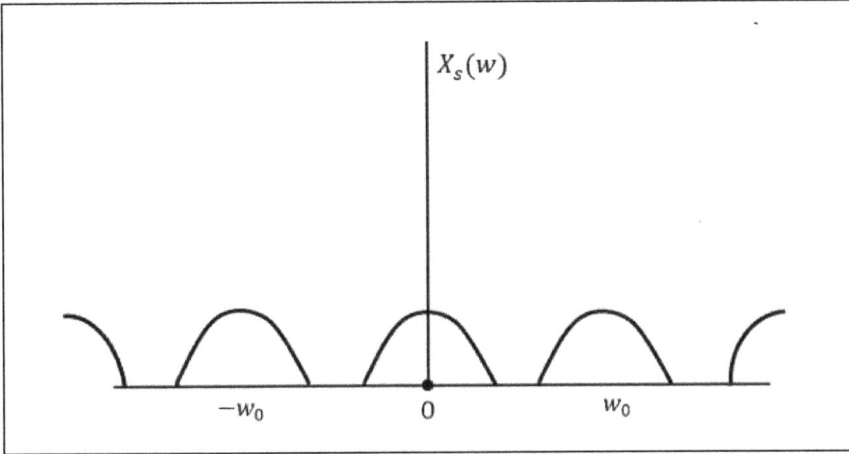

The figure shows the Fourier transform of a signal $x_s(t)$. Here, the information is reproduced without any loss. There is no mixing up and hence recovery is possible.

The Fourier Transform of the signal $x_s(t)$ is:

$$x_s(w) = \frac{1}{T_s} \sum_{n=-\infty}^{\infty} X(w - nw_0)$$

Where, T_s = Sampling Period and $w_0 = \dfrac{2\pi}{T_s}$.

Let us see what happens if the sampling rate is equal to twice the highest frequency (2W).

That means,

$$f_s = 2W$$

Where,

- f_s is the sampling frequency.

- W is the highest frequency.

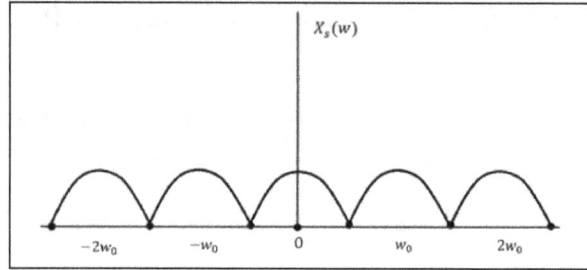

The result will be as shown in the above figure. The information is replaced without any loss. Hence, this is also a good sampling rate.

Now, let us look at the condition,

$$f_s < 2W$$

The resultant pattern will look like the following figure.

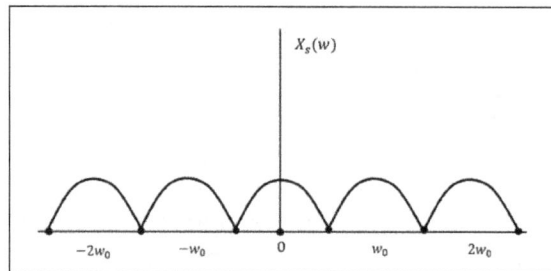

We can observe from the above pattern that the over-lapping of information is done, which leads to mixing up and loss of information. This unwanted phenomenon of over-lapping is called as Aliasing.

Scope of Fourier Transform

It is generally observed that, we seek the help of Fourier series and Fourier transforms in analyzing the signals and also in proving theorems. It is because:

- The Fourier Transform is the extension of Fourier series for non-periodic signals.

- Fourier transform is a powerful mathematical tool which helps to view the signals in different domains and helps to analyze the signals easily.

- Any signal can be decomposed in terms of sum of sines and cosines using this Fourier transform.

Aliasing

Aliasing is an effect that causes different signals to become indistinguishable from each other during sampling. Aliasing is characterized by the altering of output compared to the original signal because resampling or interpolation resulted in a lower resolution in images, a slower frame rate in terms of video or a lower wave resolution in audio. Anti-aliasing filters can be used to correct this problem.

In a digital image, aliasing manifests itself as a moiré pattern or a rippling effect. This spatial aliasing in the pattern of the image makes it look like it has waves or ripples radiating from a certain portion. This happens because the pixelation of the image is poor; when our eyes interpolate those pixels, they simply do not look right.

Aliasing can also occur in videos, where it is called temporal aliasing because it is caused by the frequency of the frames rather than the pixelation of the image. Because of the limited frame rate, a fast-moving object like a wheel looks like it's turning in reverse or too slowly; this is called the wagon-wheel effect. This is determined by the frame rate of the camera and can be avoided by using temporal aliasing reduction filters during filming.

In audio, aliasing is the result of a lower resolution sampling, which translates to poor sound quality and static. This occurs when audio is sampled at a lower resolution than the original recording. When the sinusoidal audio wave is converted to a digital wave using a lower resolution sample, only a few specific points of the wave are taken as data. This results in a wave with a lower frequency than the original, translating to a loss of data and audio quality.

Quantization

The digitization of analog signals involves the rounding off of the values which are approximately equal to the analog values. The method of sampling chooses a few points on the analog signal and then these points are joined to round off the value to a near stabilized value. Such a process is called as Quantization.

Quantizing an Analog Signal

The analog-to-digital converters perform this type of function to create a series of digital values out of the given analog signal. The following figure represents an analog signal. This signal to get converted into digital has to undergo sampling and quantizing.

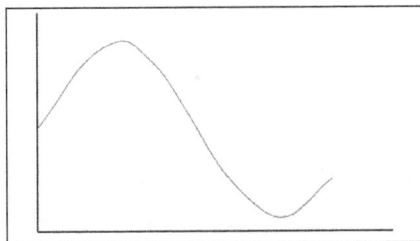

The quantizing of an analog signal is done by discretizing the signal with a number of quantization levels. Quantization is representing the sampled values of the amplitude by a finite set of levels, which means converting a continuous-amplitude sample into a discrete-time signal.

The following figure shows how an analog signal gets quantized. The blue line represents analog signal while the brown one represents the quantized signal.

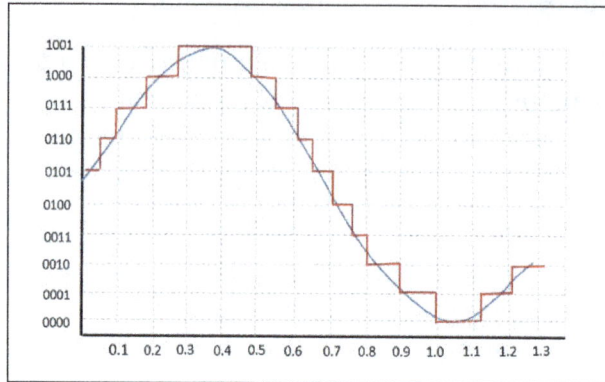

Both sampling and quantization result in the loss of information. The quality of a Quantizer output depends upon the number of quantization levels used. The discrete amplitudes of the quantized output are called as representation levels or reconstruction levels. The spacing between the two adjacent representation levels is called a quantum or step-size.

The following figure shows the resultant quantized signal which is the digital form for the given analog signal.

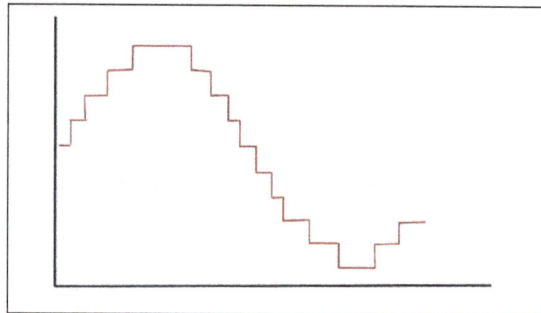

This is also called as Stair-case waveform, in accordance with its shape.

Types of Quantization

There are two types of Quantization - Uniform Quantization and Non-uniform Quantization.

The type of quantization in which the quantization levels are uniformly spaced is termed as a Uniform Quantization. The type of quantization in which the quantization levels are unequal and mostly the relation between them is logarithmic, is termed as a Non-uniform Quantization.

There are two types of uniform quantization. They are Mid-Rise type and Mid-Tread type. The following figures represent the two types of uniform quantization.

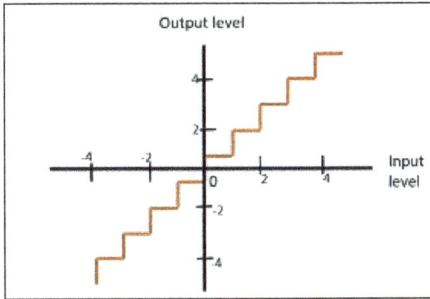

Mid-Rise type uniform quantization.

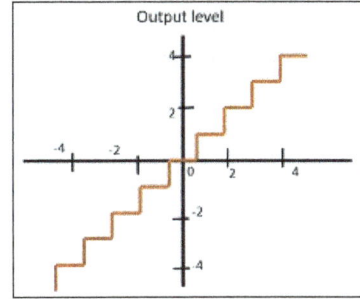

Mid-Tread type uniform quantization.

Figures show the mid-rise type and the mid-tread type of uniform quantization.

- The Mid-Rise type is so called because the origin lies in the middle of a raising part of the stair-case like graph. The quantization levels in this type are even in number.

- The Mid-tread type is so called because the origin lies in the middle of a tread of the stair-case like graph. The quantization levels in this type are odd in number.

- Both the mid-rise and mid-tread type of uniform quantizers are symmetric about the origin.

Quantization Error

For any system, during its functioning, there is always a difference in the values of its input and output. The processing of the system results in an error, which is the difference of those values.

The difference between an input value and its quantized value is called a Quantization Error. A Quantizer is a logarithmic function that performs Quantization (rounding off the value). An analog-to-digital converter (ADC) works as a quantizer.

The following figure illustrates an example for a quantization error, indicating the difference between the original signal and the quantized signal.

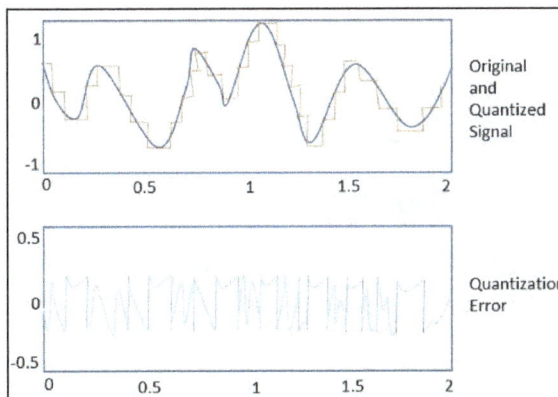

Quantization Noise

It is a type of quantization error, which usually occurs in analog audio signal, while quantizing it to digital. For example, in music, the signals keep changing continuously, where a regularity is not found in errors. Such errors create a wideband noise called as Quantization Noise.

Companding in PCM

The word Companding is a combination of Compressing and Expanding, which means that it does both. This is a non-linear technique used in PCM which compresses the data at the transmitter and expands the same data at the receiver. The effects of noise and crosstalk are reduced by using this technique.

There are two types of Companding techniques. They are:

A-law Companding Technique

- Uniform quantization is achieved at $A = 1$, where the characteristic curve is linear and no compression is done.

- A-law has mid-rise at the origin. Hence, it contains a non-zero value.

- A-law companding is used for PCM telephone systems.

μ-law Companding Technique

- Uniform quantization is achieved at $\mu = 0$, where the characteristic curve is linear and no compression is done.

- μ-law has mid-tread at the origin. Hence, it contains a zero value.

- μ-law companding is used for speech and music signals.

- μ-law is used in North America and Japan.

Sampling Techniques

Pulse Amplitude Modulation

Pulse Amplitude Modulation (PAM) is an analog modulating scheme in which the amplitude of the pulse carrier varies proportional to the instantaneous amplitude of the message signal.

The pulse amplitude modulated signal, will follow the amplitude of the original signal, as the signal traces out the path of the whole wave. In natural PAM, a signal sampled at the Nyquist rate is reconstructed, by passing it through an efficient Low Pass Frequency (LPF) with exact cutoff frequency.

The following figures explain the Pulse Amplitude Modulation.

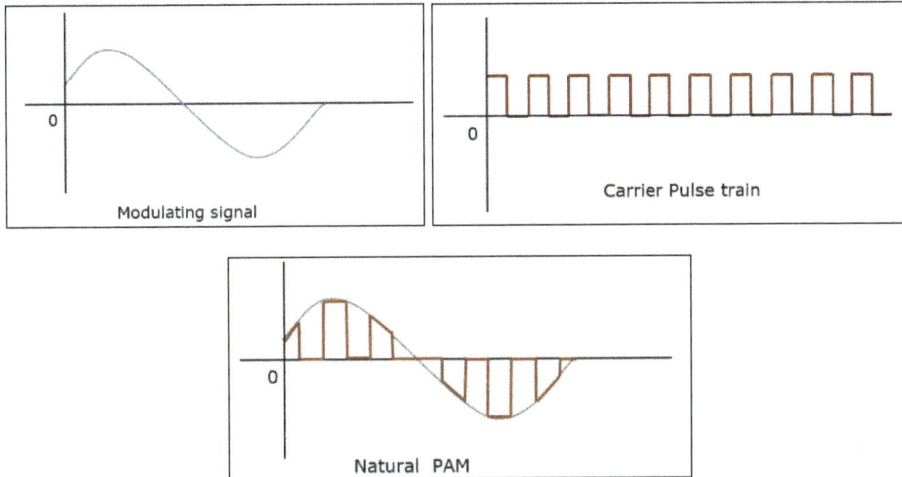

Modulating signal

Carrier Pulse train

Natural PAM

Though the PAM signal is passed through an LPF, it cannot recover the signal without distortion. Hence to avoid this noise, flat-top sampling is done as shown in the following figure.

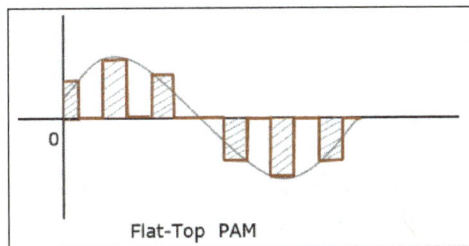

Flat-Top PAM

Flat-top sampling is the process in which sampled signal can be represented in pulses for which the amplitude of the signal cannot be changed with respect to the analog signal, to be sampled. The tops of amplitude remain flat. This process simplifies the circuit design.

Pulse Width Modulation

Pulse Width Modulation (PWM) or Pulse Duration Modulation (PDM) or Pulse Time Modulation (PTM) is an analog modulating scheme in which the duration or width or time of the pulse carrier varies proportional to the instantaneous amplitude of the message signal.

The width of the pulse varies in this method, but the amplitude of the signal remains constant. Amplitude limiters are used to make the amplitude of the signal constant. These circuits clip off the amplitude, to a desired level and hence the noise is limited.

The following figures explain the types of Pulse Width Modulations.

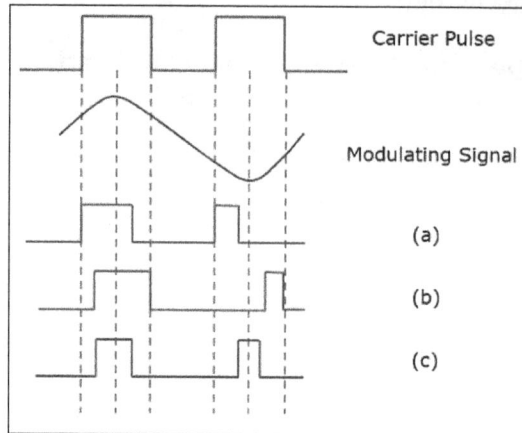

There are three variations of PWM. They are:

- The leading edge of the pulse being constant, the trailing edge varies according to the message signal.

- The trailing edge of the pulse being constant, the leading edge varies according to the message signal.

- The center of the pulse being constant, the leading edge and the trailing edge varies according to the message signal.

Pulse Position Modulation

Pulse Position Modulation (PPM) is an analog modulating scheme in which the amplitude and width of the pulses are kept constant, while the position of each pulse, with reference to the position of a reference pulse varies according to the instantaneous sampled value of the message signal.

The transmitter has to send synchronizing pulses (or simply sync pulses) to keep the transmitter and receiver in synchronism. These sync pulses help maintain the position of the pulses. The following figures explain the Pulse Position Modulation.

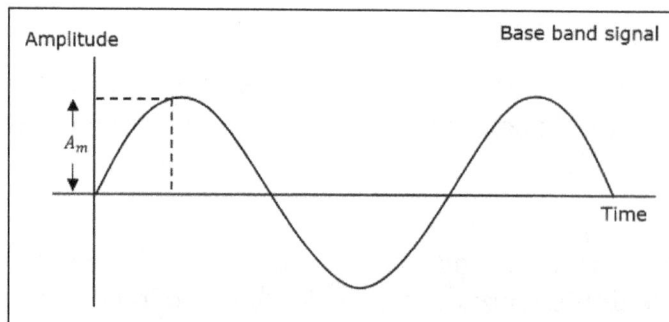

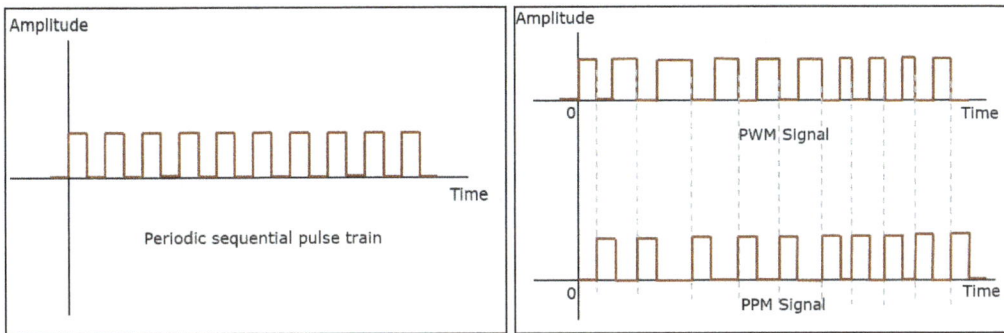

Pulse position modulation is done in accordance with the pulse width modulated signal. Each trailing of the pulse width modulated signal becomes the starting point for pulses in PPM signal. Hence, the position of these pulses is proportional to the width of the PWM pulses.

Advantage

As the amplitude and width are constant, the power handled is also constant.

Disadvantage

The synchronization between transmitter and receiver is a must.

Pulse Width Modulation

Pulse width modulation (PWM) is a powerful technique for controlling analog circuits with a microprocessor's digital outputs. PWM is employed in a wide variety of applications, ranging from measurement and communications to power control and conversion.

Analog Circuits

An analog signal has a continuously varying value, with infinite resolution in both time and magnitude. A nine-volt battery is an example of an analog device, in that its output voltage is not precisely 9V, changes over time, and can take any real-numbered value. Similarly, the amount of current drawn from a battery is not limited to a finite set of possible values. Analog signals are distinguishable from digital signals because the latter always take values only from a finite set of predetermined possibilities, such as the set {0V, 5V}.

Analog voltages and currents can be used to control things directly, like the volume of a car radio. In a simple analog radio, a knob is connected to a variable resistor. As you turn the knob, the resistance goes up or down. As that happens, the current flowing through the resistor increases or decreases. This changes the amount of current driving the speakers, thus increasing or decreasing the volume. An analog circuit is one, like the radio, whose output is linearly proportional to its input.

As intuitive and simple as analog control may seem, it is not always economically attractive or otherwise practical. For one thing, analog circuits tend to drift over time and can, therefore, be very difficult to tune. Precision analog circuits, which solve that problem, can be very large, heavy (just think of older home stereo equipment), and expensive. Analog circuits can also get very hot; the power dissipated is proportional to the voltage across the active elements multiplied by the current through them. Analog circuitry can also be sensitive to noise. Because of its infinite resolution, any perturbation or noise on an analog signal necessarily changes the current value.

Digital Control

By controlling analog circuits digitally, system costs and power consumption can be drastically reduced. What's more, many microcontrollers and DSPs already include on-chip PWM controllers, making implementation easy.

In a nutshell, PWM is a way of digitally encoding analog signal levels. Through the use of high-resolution counters, the duty cycle of a square wave is modulated to encode a specific analog signal level. The PWM signal is still digital because, at any given instant of time, the full DC supply is either fully on or fully off. The voltage or current source is supplied to the analog load by means of a repeating series of on and off pulses. The on-time is the time during which the DC supply is applied to the load, and the off-time is the period during which that supply is switched off. Given a sufficient bandwidth, any analog value can be encoded with PWM.

Figure shows three different PWM signals. figure shows a PWM output at a 10% duty cycle. That is, the signal is on for 10% of the period and off the other 90%. Figures b and c show PWM outputs at 50% and 90% duty cycles, respectively. These three PWM outputs encode three different analog signal values, at 10%, 50%, and 90% of the full strength. If, for example, the supply is 9V and the duty cycle is 10%, a 0.9V analog signal results.

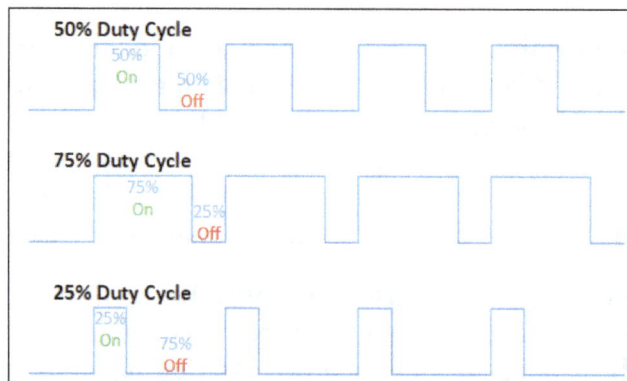

PWM signals of varying duty cycles.

Figure shows a simple circuit that could be driven using PWM. In the figure, a 9V battery powers an incandescent lightbulb. If we closed the switch connecting the battery and lamp for 50ms, the bulb would receive 9V during that interval. If we then opened the switch for the next 50ms, the bulb would receive 0V. If we repeat this cycle 10 times a second, the bulb will be lit as though it were connected to a 4.5V battery (50% of 9V). We say that the duty cycle is 50% and the modulating frequency is 10Hz.

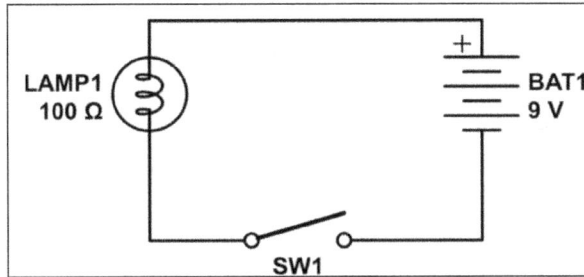

A simple circuit.

Most loads, inductive and capacitive alike, require a much higher modulating frequency than 10Hz. Imagine that our lamp was switched on for five seconds, then off for five seconds, then on again. The duty cycle would still be 50%, but the bulb would appear brightly lit for the first five seconds and off for the next. In order for the bulb to see a voltage of 4.5 volts, the cycle period must be short relative to the load's response time to a change in the switch state. To achieve the desired effect of a dimmer (but always lit) lamp, it is necessary to increase the modulating frequency. The same is true in other applications of PWM. Common modulating frequencies range from 1kHz to 200kHz.

Hardware Controllers

Many microcontrollers include PWM controllers. For example, Microchip's PIC16C67 includes two, each of which has a selectable on-time and period. The duty cycle is the ratio of the on-time to the period; the modulating frequency is the inverse of the period. To start PWM operation, the data sheet suggests the software should:

- Set the period in the on-chip timer/counter that provides the modulating square wave.

- Set the on-time in the PWM control register.

- Set the direction of the PWM output, which is one of the general-purpose I/O pins.

- Start the timer.

- Enable the PWM controller.

Although specific PWM controllers do vary in their programmatic details, the basic idea is generally the same.

Communication and Control

One of the advantages of PWM is that the signal remains digital all the way from the processor to the controlled system; no digital-to-analog conversion is necessary. By keeping the signal digital, noise effects are minimized. Noise can only affect a digital signal if it is strong enough to change a logical-1 to a logical-0, or vice versa.

Increased noise immunity is yet another benefit of choosing PWM over analog control, and is the principal reason PWM is sometimes used for communication. Switching from an analog signal to PWM can increase the length of a communications channel dramatically. At the receiving end, a suitable RC (resistor-capacitor) or LC (inductor-capacitor) network can remove the modulating high frequency square wave and return the signal to analog form.

PWM finds application in a variety of systems. As a concrete example, consider a PWM-controlled brake. To put it simply, a brake is a device that clamps down hard on something. In many brakes, the amount of clamping pressure (or stopping power) is controlled with an analog input signal. The more voltage or current that's applied to the brake, the more pressure the brake will exert.

The output of a PWM controller could be connected to a switch between the supply and the brake. To produce more stopping power, the software need only increase the duty cycle of the PWM output. If a specific amount of braking pressure is desired, measurements would need to be taken to determine the mathematical relationship between duty cycle and pressure. (And the resulting formulae or lookup tables would be tweaked for operating temperature, surface wear, and so on.)

To set the pressure on the brake to, say, 100psi, the software would do a reverse lookup to determine the duty cycle that should produce that amount of force. It would then set the PWM duty cycle to the new value and the brake would respond accordingly. If a sensor is available in the system, the duty cycle can be tweaked, under closed-loop control, until the desired pressure is precisely achieved.

Pulse Position Modulation

Pulse position modulation is a signal modulation technique that allows computers to share data by measuring the time each data packet takes to reach the computer. It is often used in optical communication, such as fiber optics, in which there is little multi-pathway interference. Pulse position modulation exclusively transfers digital signals and cannot be used with analog systems. It transfers simple data and is not effective at transferring files.

How Pulse Position Modulation Works

Pulse position modulation works by sending electrical, electromagnetic, or optical pulses to a computer or other device in order to communicate simple data. It requires both devices to be synchronized to the same clock so that when a series of pulses is sent, the device decodes the information based on when the pulses were broadcasted. Alternately, another form of pulse position modulation known as differential pulse position modulation, allows all signals to be encoded based on the difference between broadcast times. This means that a receiving device only has to observe the difference in arrival times in order to decode a transmission.

Applications

Pulse position modulation has many purposes, especially in RF (Radio Frequency) communications. For example, pulse position modulation is used in remote controlled aircraft, cars, boats, and other vehicles and is responsible for conveying a transmitter's controls to a receiver. Each pulse's position may describe an analogue controller's physical direction, while the number of pulses may describe the number of possible commands that the device may receive.

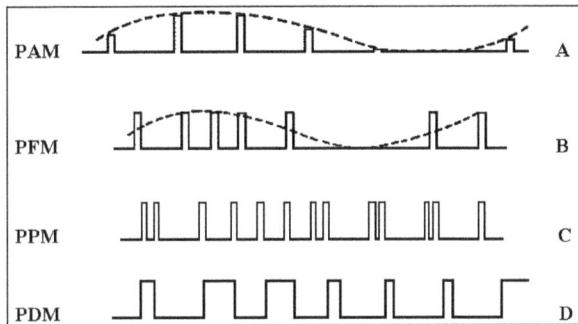

Advantages

Pulse position modulation conveys simple commands that other forms of signal modulation are either simply not made for or are too complex to use in certain situations. Because pulse position modulation only communicates simple commands from a transmitter to a receiver, it is often used in lightweight applications due to its low system requirements.

Disadvantages

Pulse position modulation requires that both devices are synchronized or differential pulse position modulation is used. Also, pulse position modulation is highly sensitive to multi-pathway interference, such as echoing, that can disrupt a transmission by altering the difference in arrival times of each signal.

Pulse Code Modulation

There are many modulation techniques, which are classified according to the type of modulation employed. Of them all, the digital modulation technique used is Pulse Code Modulation (PCM).

A signal is pulse code modulated to convert its analog information into a binary sequence, i.e., 1s and 0s. The output of a PCM will resemble a binary sequence. The following figure shows an example of PCM output with respect to instantaneous values of a given sine wave.

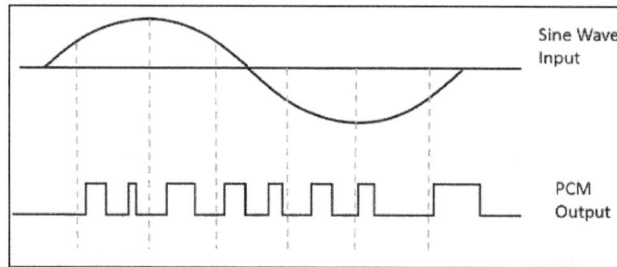

Instead of a pulse train, PCM produces a series of numbers or digits, and hence this process is called as digital. Each one of these digits, though in binary code, represent the approximate amplitude of the signal sample at that instant.

In Pulse Code Modulation, the message signal is represented by a sequence of coded pulses. This message signal is achieved by representing the signal in discrete form in both time and amplitude.

Basic Elements of PCM

The transmitter section of a Pulse Code Modulator circuit consists of Sampling, Quantizing and Encoding, which are performed in the analog-to-digital converter section. The low pass filter prior to sampling prevents aliasing of the message signal.

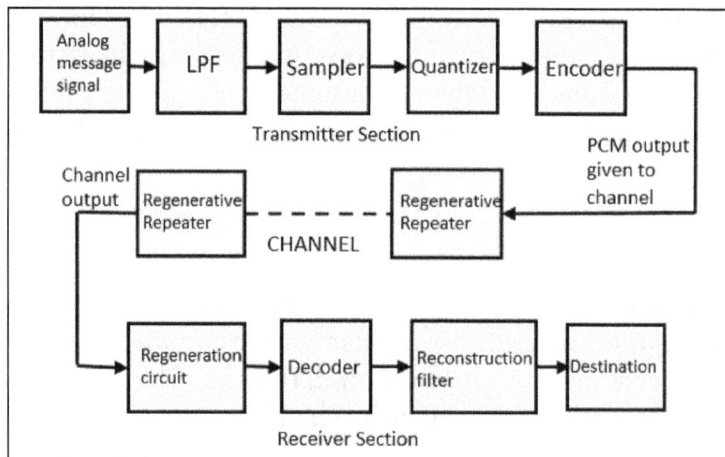

The basic operations in the receiver section are regeneration of impaired signals, decoding, and reconstruction of the quantized pulse train. Following is the block diagram of PCM which represents the basic elements of both the transmitter and the receiver sections.

Low Pass Filter

This filter eliminates the high frequency components present in the input analog signal which is greater than the highest frequency of the message signal, to avoid aliasing of the message signal.

Sampler

This is the technique which helps to collect the sample data at instantaneous values of message signal, so as to reconstruct the original signal. The sampling rate must be greater than twice the highest frequency component W of the message signal, in accordance with the sampling theorem.

Quantizer

Quantizing is a process of reducing the excessive bits and confining the data. The sampled output when given to Quantizer, reduces the redundant bits and compresses the value.

Encoder

The digitization of analog signal is done by the encoder. It designates each quantized level by a binary code. The sampling done here is the sample-and-hold process. These three sections (LPF, Sampler, and Quantizer) will act as an analog to digital converter. Encoding minimizes the bandwidth used.

Regenerative Repeater

This section increases the signal strength. The output of the channel also has one regenerative repeater circuit, to compensate the signal loss and reconstruct the signal, and also to increase its strength.

Decoder

The decoder circuit decodes the pulse coded waveform to reproduce the original signal. This circuit acts as the demodulator.

Reconstruction Filter

After the digital-to-analog conversion is done by the regenerative circuit and the decoder, a low-pass filter is employed, called as the reconstruction filter to get back the original signal.

Hence, the Pulse Code Modulator circuit digitizes the given analog signal, codes it and samples it, and then transmits it in an analog form. This whole process is repeated in a reverse pattern to obtain the original signal.

Digital Modulation Techniques

Amplitude Shift Keying

Amplitude Shift Keying (ASK) is a type of Amplitude Modulation which represents the binary data in the form of variations in the amplitude of a signal.

Any modulated signal has a high frequency carrier. The binary signal when ASK modulated, gives a zero value for Low input while it gives the carrier output for High input.

The following figure represents ASK modulated waveform along with its input.

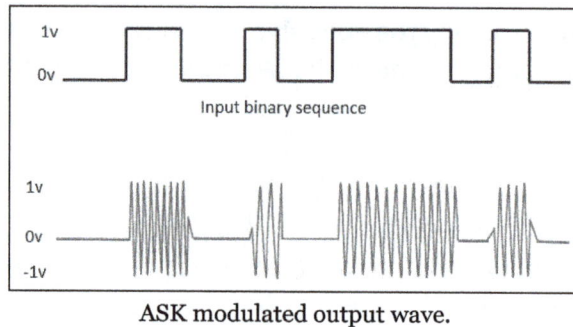

ASK modulated output wave.

To find the process of obtaining this ASK modulated wave, let us learn about the working of the ASK modulator.

ASK Modulator

The ASK modulator block diagram comprises of the carrier signal generator, the binary sequence from the message signal and the band-limited filter. Following is the block diagram of the ASK Modulator.

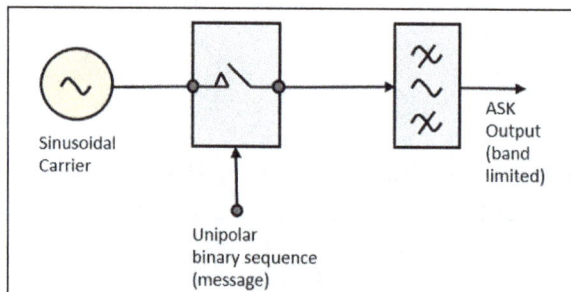

ASK Generation method.

The carrier generator, sends a continuous high-frequency carrier. The binary sequence from the message signal makes the unipolar input to be either High or Low. The high signal closes the switch, allowing a carrier wave. Hence, the output will be the carrier signal at high input. When there is low input, the switch opens, allowing no voltage to appear. Hence, the output will be low.

The band-limiting filter, shapes the pulse depending upon the amplitude and phase characteristics of the band-limiting filter or the pulse-shaping filter.

ASK Demodulator

There are two types of ASK Demodulation techniques. They are:

- Asynchronous ASK Demodulation/detection.

- Synchronous ASK Demodulation/detection.

The clock frequency at the transmitter when matches with the clock frequency at the receiver, it is known as a Synchronous method, as the frequency gets synchronized. Otherwise, it is known as Asynchronous.

Asynchronous ASK Demodulator

The Asynchronous ASK detector consists of a half-wave rectifier, a low pass filter, and a comparator. Following is the block diagram for the same.

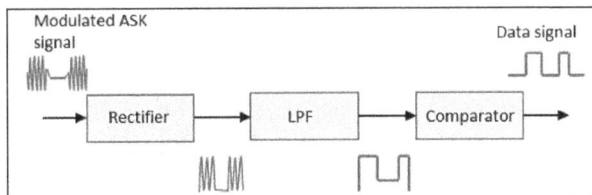

Asynchronous ASK detector.

The modulated ASK signal is given to the half-wave rectifier, which delivers a positive half output. The low pass filter suppresses the higher frequencies and gives an envelope detected output from which the comparator delivers a digital output.

Synchronous ASK Demodulator

Synchronous ASK detector consists of a Square law detector, low pass filter, a comparator, and a voltage limiter. Following is the block diagram for the same.

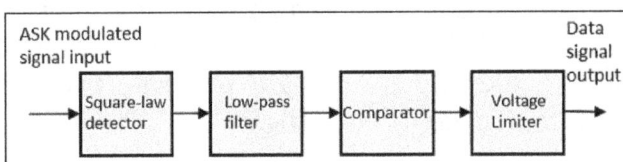

Synchronous ASK detector.

The ASK modulated input signal is given to the Square law detector. A square law detector is one whose output voltage is proportional to the square of the amplitude modulated input voltage. The low pass filter minimizes the higher frequencies. The comparator and the voltage limiter help to get a clean digital output.

Frequency Shift Keying

Frequency Shift Keying (FSK) is the digital modulation technique where the frequency of the carrier signal varies as per the changes in the digital signal.

For a binary high input, the output of a FSK modulated wave is high in frequency and for binary low input its low in frequency. The binary 0s and 1s are called as Space and Mark frequencies.

Figure below is the FSK modulated waveform with its input.

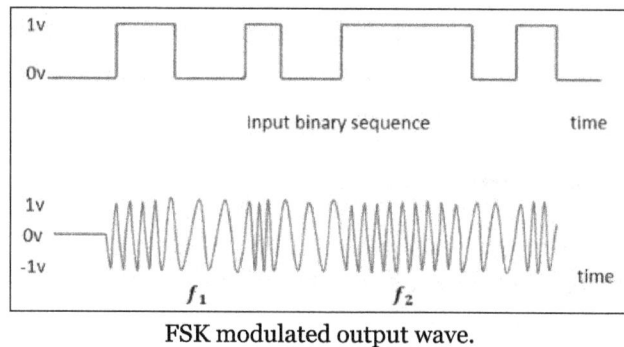

FSK modulated output wave.

FSK Modulator

It contains two oscillators with a clock and the input binary sequence.

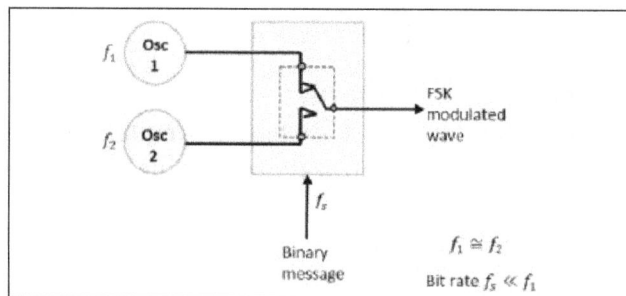

FSK Transmitter.

Two oscillators each producing a lower and higher frequency signals are connected to a switch along with an internal clock. This clock is applied to both oscillators to avoid the phase discontinuities of the output waveform during the transmission of the signal. The binary input sequence is given to the transmitter to choose the frequencies based on the binary input.

FSK Demodulator

Several methods for demodulating a FSK wave are available and the main methods for detecting are asynchronous detector which is non-coherent and synchronous detector which is coherent.

Asynchronous FSK Detector

The Asynchronous FDK Detector consists of two band pass filters, two envelope detectors and a decision circuit.

The FSK signal is given to the detector and two Band Pass Filters tuning to Space and Mark frequencies. The output looks like ASK signal which is headed to envelope detector. The signal is modulated in each envelope detector.

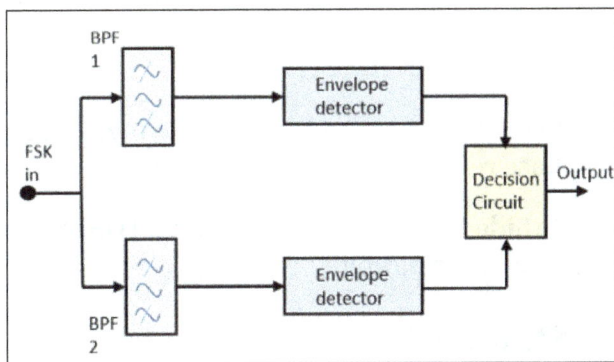

Decision circuit will selects the output which is more likely and re-shapes the waveform to a rectangular one.

Synchronous FSK Detector

Block diagram contains two mixers with local oscillator circuits, two band pass filters and a decision circuit.

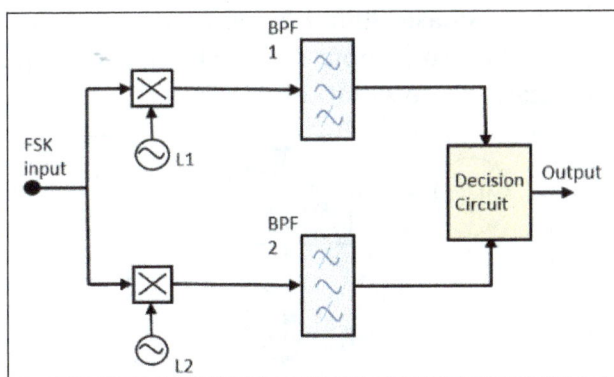

The FSK signal is given as input to the two mixers with local oscillator circuits which are connected to two BPF. This combination acts as demodulators and the decision circuit decides in choosing the best output from the detectors. The two signals have minimum separation in frequency.

Bandwidth of each of the demodulators depends on their bit rate. Synchronous demodulator is bit complex than Asynchronous demodulator.

Phase Shift Keying

Phase Shift Keying (PSK) is the digital modulation technique in which the phase of the carrier signal is changed by varying the sine and cosine inputs at a particular time. PSK technique is widely used for wireless LANs, bio-metric, contactless operations, along with RFID and Bluetooth communications.

PSK is of two types, depending upon the phases the signal gets shifted. They are:

Binary Phase Shift Keying

This is also called as 2-phase PSK or Phase Reversal Keying. In this technique, the sine wave carrier takes two phase reversals such as 0° and 180°.

BPSK is basically a Double Side Band Suppressed Carrier (DSBSC) modulation scheme, for message being the digital information.

Quadrature Phase Shift Keying

This is the phase shift keying technique, in which the sine wave carrier takes four phase reversals such as 0°, 90°, 180°, and 270°.

If this kind of techniques is further extended, PSK can be done by eight or sixteen values also, depending upon the requirement.

BPSK Modulator

The block diagram of Binary Phase Shift Keying consists of the balance modulator which has the carrier sine wave as one input and the binary sequence as the other input. Following is the diagrammatic representation.

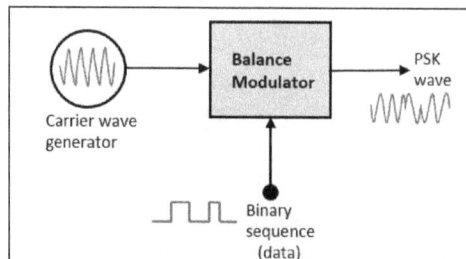

The modulation of BPSK is done using a balance modulator, which multiplies the two signals applied at the input. For a zero binary input, the phase will be 0° and for a high input, the phase reversal is of 180°.

Following is the diagrammatic representation of BPSK Modulated output wave along with its given input.

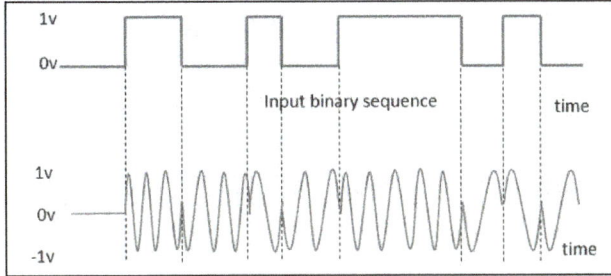

BPSK Modulator output wave.

The output sine wave of the modulator will be the direct input carrier or the inverted (180° phase shifted) input carrier, which is a function of the data signal.

BPSK Demodulator

The block diagram of BPSK demodulator consists of a mixer with local oscillator circuit, a bandpass filter, a two-input detector circuit. The diagram is as follows.

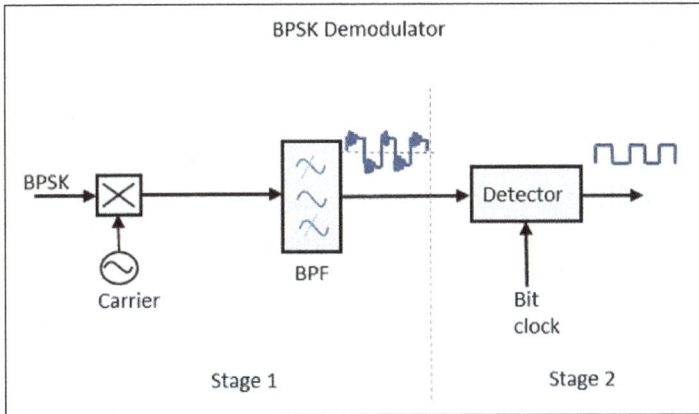

By recovering the band-limited message signal, with the help of the mixer circuit and the band pass filter, the first stage of demodulation gets completed. The base band signal which is band limited is obtained and this signal is used to regenerate the binary message bit stream.

In the next stage of demodulation, the bit clock rate is needed at the detector circuit to produce the original binary message signal. If the bit rate is a sub-multiple of the carrier frequency, then the bit clock regeneration is simplified. To make the circuit easily understandable, a decision-making circuit may also be inserted at the 2nd stage of detection.

Quadrature Phase Shift Keying

The Quadrature Phase Shift Keying (QPSK) is one type of BPSK and also a Double Side Band Suppressed Carrier (DSBSC) modulation scheme that transmits two bits of digital information at an instant called as bigits.

Instead of converting the digital bits to a series of digital stream, they are converted to bit pairs, which decrease the bit rate to half allowing space for the other users.

QPSK Modulator

The QPSK Modulator employs a bit-splitter, two multipliers along with local oscillator, a 2-bit serial to parallel converter and a summer circuit.

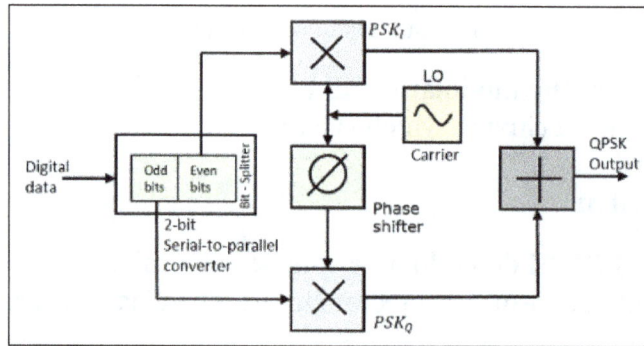

Input to the modulator is the message signal and the bit splitter separates even and odd bits and are multiplied with the same carrier to produce odd BPS_K (called as PSK_I) and even BPS_K (called as PSK_Q).

Modulated output for different instances of binary inputs is shown below.

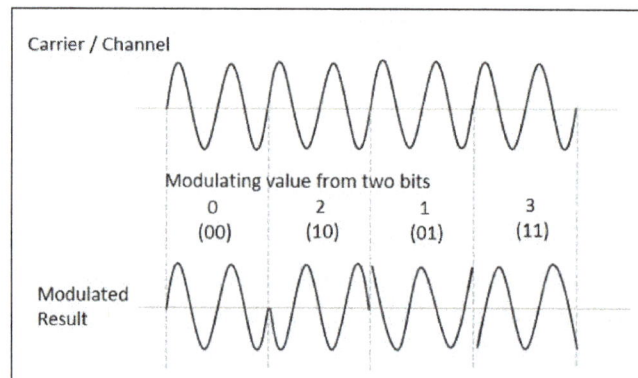

QPSK Demodulator

It employs two product demodulator circuits along with local oscillator, two BPF, two integrator circuits and a 2-bit parallel to serial converter.

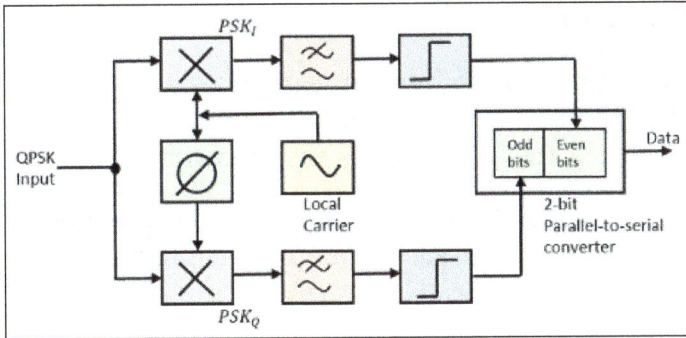

The two BPSK signals are demodulated by the two product detectors at the input. From the original data, the pair of bits is recovered. These signals are passed to the parallel to serial converter after processing.

M-ary Encoding

The word binary represents two bits. M represents a digit that corresponds to the number of conditions, levels, or combinations possible for a given number of binary variables.

This is the type of digital modulation technique used for data transmission in which instead of one bit, two or more bits are transmitted at a time. As a single signal is used for multiple bit transmission, the channel bandwidth is reduced.

M-ary Equation

If a digital signal is given under four conditions, such as voltage levels, frequencies, phases, and amplitude, then M = 4.

The number of bits necessary to produce a given number of conditions is expressed mathematically as,

$$N = \log_2 M$$

where,

N is the number of bits necessary.

M is the number of conditions, levels, or combinations possible with N bits.

The above equation can be re-arranged as,

$$2^N = M$$

For example, with two bits, $2^2 = 4$ conditions are possible.

Types of M-ary Techniques

In general, Multi-level (M-ary) modulation techniques are used in digital communications as the digital inputs with more than two modulation levels are allowed on the transmitter's input. Hence, these techniques are bandwidth efficient.

There are many M-ary modulation techniques. Some of these techniques, modulate one parameter of the carrier signal, such as amplitude, phase, and frequency.

M-ary ASK

This is called M-ary Amplitude Shift Keying (M-ASK) or M-ary Pulse Amplitude Modulation (PAM).

The amplitude of the carrier signal, takes on M different levels.

Representation of M-ary ASK

$$S_m(t) = A_m \cos(2\pi f_c t) \quad A_m \varepsilon (2M-1-M)\Delta, m=1,2.....M \text{ and } 0 \le t \le T_s$$

Some prominent features of M-ary ASK are:

- This method is also used in PAM.
- Its implementation is simple.
- M-ary ASK is susceptible to noise and distortion.

M-ary FSK

This is called as M-ary Frequency Shift Keying (M-ary FSK). The frequency of the carrier signal, takes on M different levels.

Representation of M-ary FSK

$$S_i(t) = \sqrt{\frac{2E_s}{T_s}} \cos\left(\frac{\pi}{T_s}(n_c+i)t\right); \ 0 \le t \le T_s \text{ and } i=1,2,3...M$$

Where,

$$f_c = \frac{n_c}{2T_s} \text{ for some fixed integer n.}$$

Some prominent features of M-ary FSK are:

- Not susceptible to noise as much as ASK.
- The transmitted M number of signals are equal in energy and duration.

- The signals are separated by $\dfrac{1}{2T_s}$ Hz making the signals orthogonal to each other.

- Since M signals are orthogonal, there is no crowding in the signal space.

- The bandwidth efficiency of M-ary FSK decreases and the power efficiency increases with the increase in M.

M-ary PSK

This is called as M-ary Phase Shift Keying (M-ary PSK). The phase of the carrier signal, takes on M different levels.

Representation of M-ary PSK

$$S_i(t) = \sqrt{\dfrac{2E}{T}}\cos(w_o t + \phi_i t);\ 0 \le t \le T \quad \text{and} \quad i = 1,2....M$$

$$\phi_i(t) = \dfrac{2\pi i}{M}\ where\ i = 1,2,3.... \ ... M$$

Some prominent features of M-ary PSK are:

- The envelope is constant with more phase possibilities.

- This method was used during the early days of space communication.

- Better performance than ASK and FSK.

- Minimal phase estimation error at the receiver.

- The bandwidth efficiency of M-ary PSK decreases and the power efficiency increases with the increase in M.

The output of all these techniques is a binary sequence, represented as 1s and 0s. This binary or digital information has many types and forms.

References

- Digital-communication-quick-guide: tutorialspoint.com, Retrieved 19 April, 2019
- Digital-communication-sampling: tutorialspoint.com, Retrieved 23 August, 2019
- Digital-communication-quantization: torialspoint.com, Retrieved 25 May, 2019
- Principles-of-communication-analog-pulse-modulation: tutorialspoint.com, Retrieved 15 January, 2019
- Introduction-to-Pulse-Width-Modulation: embedded.com, Retrieved 19 February, 2019
- Pulse-position-modulation: tech-faq.com, Retrieved 9 March, 2019

- Digital-communication-pulse-code-modulation, digital-communication: tutorialspoint.com, Retrieved 28 February, 2019

- Digital-communication-amplitude-shift-keying, digital-communication: tutorialspoint.com, Retrieved 30 April, 2019

- Frequency-shift-keying-25993: wisdomjobs.com, Retrieved 1 March, 2019

- Digital-communication-phase-shift-keying, digital-communication: tutorialspoint.com, Retrieved 14 June, 2019

- Quadrature-phase-shift-keying-25996: wisdomjobs.com, Retrieved 11 January, 2019

- Digital-communication-m-ary-encoding, digital-communication: tutorialspoint.com, Retrieved 8 July, 2019

4

Wireless Networking

Wireless networking is a method through which different network nodes are connected through wireless data connections. Some of the different types of wireless networks are personal area networks, global area networks and cellular wireless networks. The topics elaborated in this chapter will help in gaining a better perspective about these types of wireless networks.

Satellite Communications

A satellite is an object that revolves around another object. For example, earth is a satellite of The Sun, and moon is a satellite of earth.

A communication satellite is a microwave repeater station in a space that is used for telecommunication, radio and television signals. A communication satellite processes the data coming from one earth station and it converts the data into another form and sends it to the second earth station.

Working of Satellite

Two stations on earth want to communicate through radio broadcast but are too far away to use conventional means. The two stations can use a relay station for their communication. One earth station transmits the signal to the satellite.

Uplink frequency is the frequency at which ground station is communicating with satellite. The satellite transponder converts the signal and sends it down to the second earth station, and this is called Downlink frequency. The second earth station also communicates with the first one in the same way.

Advantages of Satellite

The advantages of Satellite Communications are as follows:

- The Coverage area is very high than that of terrestrial systems.
- The transmission cost is independent of the coverage area.
- Higher bandwidths are possible.

Disadvantages of Satellite

The disadvantages of Satellite Communications are as follows:

- Launching satellites into orbits is a costly process.

- The bandwidths are gradually used up.

- High propagation delay for satellite systems than the conventional terrestrial systems.

The process of satellite communication begins at an earth station. Here an installation is designed to transmit and receive signals from a satellite in orbit around the earth. Earth stations send information to satellites in the form of high powered, high frequency (GHz range) signals.

The satellites receive and retransmit the signals back to earth where they are received by other earth stations in the coverage area of the satellite. Satellite's footprint is the area which receives a signal of useful strength from the satellite.

The transmission system from the earth station to the satellite through a channel is called the uplink. The system from the satellite to the earth station through the channel is called the downlink.

Satellite Frequency Bands

The satellite frequency bands which are commonly used for communication are the Cb and, Ku-band, and Ka-band. C-band and Ku-band are the commonly used frequency spectrums by today's satellites.

It is important to note that there is an inverse relationship between frequency and wavelength i.e. when frequency increases, wavelength decreases this helps to understand the relationship between antenna diameter and transmission frequency. Larger antennas (satellite dishes) are necessary to gather the signal with increasing wavelength.

Earth Orbits

A satellite when launched into space needs to be placed in certain orbit to provide a particular way for its revolution, so as to maintain accessibility and serve its purpose whether scientific, military or commercial. Such orbits which are assigned to satellites, with respect to earth are called as Earth Orbits. The satellites in these orbits are Earth Orbit Satellites.

The important kinds of Earth Orbits are:

- Geo-synchronous Earth Orbit,

- Geo-stationary Earth Orbit,

- Medium Earth Orbit,

- Low Earth Orbit.

Geo-synchronous Earth Orbit Satellites

A Geo-synchronous Earth orbit Satellite is one which is placed at an altitude of 22,300 miles above the Earth. This orbit is synchronized with a side real day (i.e., 23 hours 56 minutes). This orbit can have inclination and eccentricity. It may not be circular. This orbit can be tilted at the poles of the earth. But it appears stationary when observed from the Earth.

The same geo-synchronous orbit, if it is circular and in the plane of equator, it is called as geo-stationary orbit. These Satellites are placed at 35,900 kms (same as geosynchronous) above the Earth's Equator and they keep on rotating with respect to earth's direction (west to east). These satellites are considered stationary with respect to earth and hence the name implies.

Geo-stationary Earth Orbit Satellites are used for weather forecasting, satellite TV, satellite radio and other types of global communications.

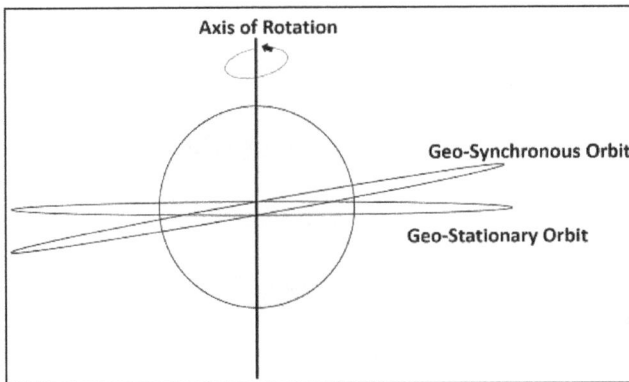

The above figure shows the difference between Geo-synchronous and Geo-stationary orbits. The Axis of rotation indicates the movement of Earth.

The main point to note here is that every Geo-stationary orbit is a Geo-synchronous orbit. But every Geo-synchronous orbit is not a Geo-stationary orbit.

Medium Earth Orbit Satellites

Medium earth orbit (MEO) satellite networks will orbit at distances of about 8000 miles from earth's surface. Signals transmitted from a MEO satellite travel a shorter distance. This translates to improved signal strength at the receiving end. This shows that smaller, more lightweight receiving terminals can be used at the receiving end.

Since the signal is travelling a shorter distance to and from the satellite, there is less transmission delay. Transmission delay can be defined as the time it takes for a signal to travel up to a satellite and back down to a receiving station.

For real-time communications, the shorter the transmission delay, the better will be the communication system. As an example, if a GEO satellite requires 0.25 seconds for a round trip, then MEO satellite requires less than 0.1 seconds to complete the same trip. MEOs operates in the frequency range of 2 GHz and above.

Low Earth Orbit Satellites

The Low Earth Orbit (LEO) satellites are mainly classified into three categories namely, little LEOs, big LEOs, and Mega-LEOs. LEOs will orbit at a distance of 500 to 1000 miles above the earth's surface.

This relatively short distance reduces transmission delay to only 0.05 seconds. This further reduces the need for sensitive and bulky receiving equipment. Little LEOs will operate in the 800 MHz (0.8 GHz) range. Big LEOs will operate in the 2 GHz or above range, and Mega-LEOs operate in the 20-30 GHz range.

The higher frequencies associated with Mega-LEOs translates into more information carrying capacity and yields to the capability of real-time, low delay video transmission scheme.

High Altitude Long Endurance Platforms

Experimental High Altitude Long Endurance (HALE) platforms are basically highly efficient and lightweight airplanes carrying communications equipment. This will act as very low earth orbit geosynchronous satellites.

These crafts will be powered by a combination of battery and solar power or high efficiency turbine engines. HALE platforms will offer transmission delays of less than 0.001 seconds at an altitude of only 70,000 feet, and even better signal strength for very lightweight hand-held receiving devices.

Orbital Slots

Here there may arise a question that with more than 200 satellites up there in geosynchronous orbit, how do we keep them from running into each other or from attempting to use the same location in space? To answer this problem, international regulatory bodies like the International Telecommunications Union (ITU) and national government organizations like the Federal Communications Commission (FCC) designate the locations on the geosynchronous orbit where the communications satellites can be located.

These locations are specified in degrees of longitude and are called as orbital slots. The FCC and ITU have progressively reduced the required spacing down to only 2 degrees for C-band and Ku-band satellites due to the huge demand for orbital slots.

Principles of Satellite Communications

A satellite is a body that moves around another body in a mathematically predictable path called an Orbit. A communication satellite is nothing but a microwave repeater station in space that is helpful in telecommunications, radio, and television along with internet applications.

A repeater is a circuit which increases the strength of the signal it receives and retransmits it. But here this repeater works as a transponder, which changes the frequency band of the transmitted signal, from the received one.

The frequency with which the signal is sent into the space is called Uplink frequency, while the frequency with which it is sent by the transponder is Downlink frequency.

The following figure illustrates this concept clearly.

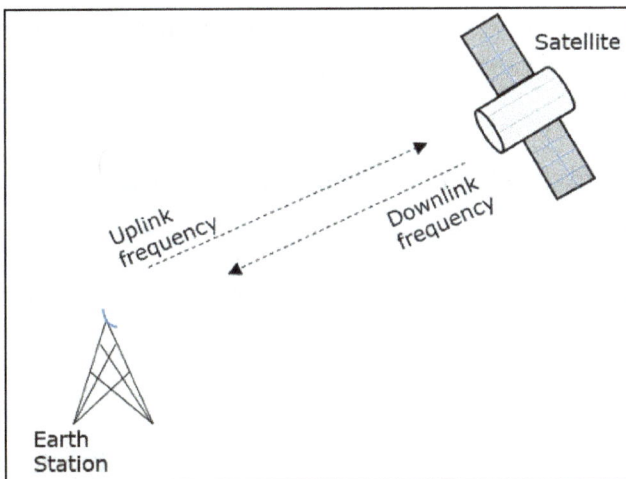

Now, let us have a look at the advantages, disadvantages and applications of satellite communications.

Advantages

There are many Advantages of satellite communications such as:

- Flexibility,
- Ease in installing new circuits,
- Distances are easily covered and cost doesn't matter,
- Broadcasting possibilities,
- Each and every corner of earth is covered,
- User can control the network.

Disadvantages

Satellite communication has the following drawbacks:

- The initial costs such as segment and launch costs are too high.
- Congestion of frequencies.
- Interference and propagation.

Applications

Satellite communication finds its applications in the following areas:

- In Radio broadcasting.
- In TV broadcasting such as DTH.
- In Internet applications such as providing Internet connection for data transfer, GPS applications, Internet surfing, etc.
- For voice communications.
- For research and development sector, in many areas.
- In military applications and navigations.

The orientation of the satellite in its orbit depends upon the three laws called as Kepler's laws.

Kepler's Laws

Johannes Kepler the astronomical scientist, gave 3 revolutionary laws, regarding the motion of satellites. The path followed by a satellite around its primary (the earth) is an ellipse. Ellipse has two foci - F1 and F2, the earth being one of them.

If the distance from the center of the object to a point on its elliptical path is considered, then the farthest point of an ellipse from the center is called as apogeeand the shortest point of an ellipse from the center is called as perigee.

Kepler's 1st Law

Kepler's 1st law states that, "every planet revolves around the sun in an elliptical orbit, with sun as one of its foci". As such, a satellite moves in an elliptical path with earth as one of its foci.

The semi major axis of the ellipse is denoted as 'a' and semi minor axis is denoted as 'b'. Therefore, the eccentricity 'e' of this system can be written as:

$$e = \frac{\sqrt{a^2 - b^2}}{a}$$

- Eccentricity (e) – It is the parameter which defines the difference in the shape of the ellipse rather than that of a circle.

- Semi-major axis (a) – It is the longest diameter drawn joining the two foci along the center, which touches both the apogees (farthest points of an ellipse from the center).

- Semi-minor axis (b) – It is the shortest diameter drawn through the center which touches both the perigees (shortest points of an ellipse from the center).

These are well described in the following figure:

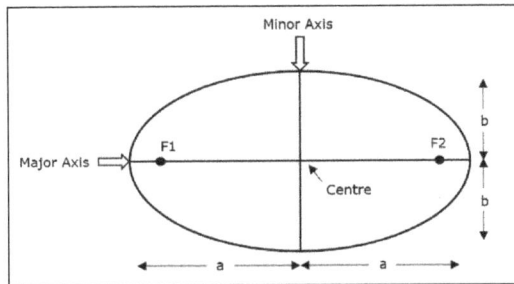

Kepler's Laws.

For an elliptical path, it is always desirable that the eccentricity should lie in between 0 and 1, i.e. $0 < e < 1$ because if e becomes zero, the path will be no more in elliptical shape rather it will be converted into a circular path.

Kepler's 2nd Law

Kepler's 2nd law states that, "For equal intervals of time, the area covered by the satellite is equal with respect to the center of the earth".

It can be understood by taking a look at the following figure:

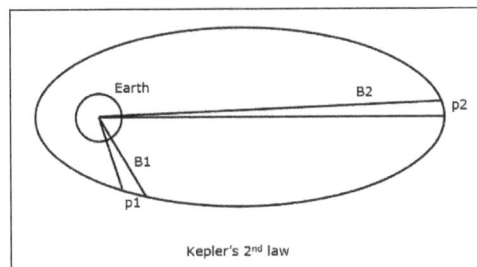

Suppose that the satellite covers p1 and p2 distances, in the same time interval, then the areas B1 and B2 covered in both instances respectively, are equal.

Kepler's 3rd Law

Kepler's 3rd law states that, "The square of the periodic time of the orbit is proportional to the cube of the mean distance between the two bodies".

This can be written mathematically as,

$$T^2 \; \alpha \; a^3$$

which implies,

$$T^2 = \frac{4\pi^2}{GM} a^3$$

where, $\dfrac{4\pi^2}{GM}$ is the proportionality constant (according to Newtonian Mechanics),

$$T^2 = \frac{4\pi^2}{\mu} a^3$$

where, μ = the earth's geocentric gravitational constant, i.e. M = 3.986005 × 10^{14} m³/sec²,

$$1 = \left(\frac{2\pi}{T}\right)^2 \frac{a^3}{\mu}$$

$$1 = n^2 \frac{a^3}{\mu} \Rightarrow a^3 = \frac{\mu}{n^2}$$

where, n = the mean motion of the satellite in radians per second.

The orbital functioning of satellites is calculated with the help of these Kepler's laws.

Along with these, there is an important thing which has to be noted. A satellite, when it revolves around the earth, undergoes a pulling force from the earth which is the gravitational force. Also, it experiences some pulling force from the sun and the moon. Hence, there are two forces acting on it. They are:

- Centripetal force – The force that tends to draw an object moving in a trajectory path, towards itself is called as centripetal force.

- Centrifugal force – The force that tends to push an object moving in a trajectory path, away from its position is called as centrifugal force.

So, a satellite has to balance these two forces to keep itself in its orbit.

Earth Orbits

A satellite when launched into space, needs to be placed in a certain orbit to provide a particular way for its revolution, so as to maintain accessibility and serve its purpose whether scientific, military, or commercial. Such orbits which are assigned to satellites, with respect to earth are called as Earth Orbits. The satellites in these orbits are Earth Orbit Satellites.

The important kinds of Earth Orbits are:

- Geo-synchronous Earth Orbit.
- Medium Earth Orbit.
- Low Earth Orbit.

Geosynchronous Earth Orbit Satellites

A Geo-synchronous Earth Orbit (GEO) satellite is one which is placed at an altitude of 22,300 miles above the Earth. This orbit is synchronized with a side real day (i.e., 23 hours 56 minutes). This orbit can have inclination and eccentricity. It may not be circular. This orbit can be tilted at the poles of the earth. But it appears stationary when observed from the Earth.

The same geo-synchronous orbit, if it is circular and in the plane of equator, it is called as geo-stationary orbit. These satellites are placed at 35,900 kms (same as geosynchronous) above the Earth's Equator and they keep on rotating with respect to earth's direction (west to east). These satellites are considered stationary with respect to earth and hence the name implies.

Geo-Stationary Earth Orbit Satellites are used for weather forecasting, satellite TV, satellite radio and other types of global communications.

The following figure shows the difference between Geo-synchronous and Geo-stationary orbits. The axis of rotation indicates the movement of Earth.

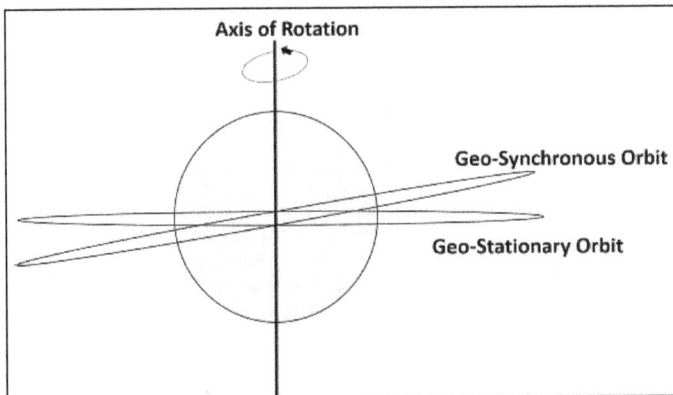

Medium Earth Orbit Satellites

Medium Earth Orbit (MEO) satellite networks will orbit at distances of about 8000 miles from the earth's surface. Signals transmitted from a MEO satellite travel a shorter distance. This translates to improved signal strength at the receiving end. This shows that smaller, more lightweight receiving terminals can be used at the receiving end.

Since the signal is travelling a shorter distance to and from the satellite, there is less transmission delay. Transmission delay can be defined as the time it takes for a signal to travel up to a satellite and back down to a receiving station.

For real-time communications, the shorter the transmission delay, the better will be the communication system. As an example, if a GEO satellite requires 0.25 seconds for a round trip, then MEO satellite requires less than 0.1 seconds to complete the same trip. MEOs operates in the frequency range of 2 GHz and above.

Low Earth Orbit Satellites

The Low Earth Orbit (LEO) satellites are mainly classified into three categories namely, little LEOs, big LEOs, and Mega-LEOs. LEOs will orbit at a distance of 500 to 1000 miles above the earth's surface.

This relatively short distance reduces transmission delay to only 0.05 seconds. This further reduces the need for sensitive and bulky receiving equipment. Little LEOs will operate in the 800 MHz (0.8 GHz) range. Big LEOs will operate in the 2 GHz or above range, and Mega-LEOs operates in the 20-30 GHz range.

The higher frequencies associated with Mega-LEOs translates into more information carrying capacity and yields to the capability of real-time, low delay video transmission scheme.

The following figure depicts the paths of LEO, MEO, and GEO.

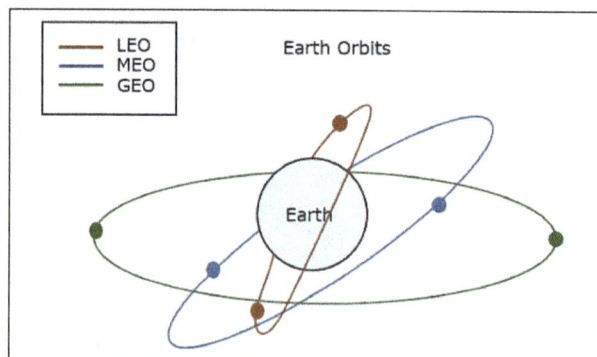

Wireless Distribution System

A wireless distribution system (WDS) is a method of interconnecting access points (AP) in a wireless local area network (WLAN) without requiring that they connect through a wired backbone.

The IEEE 802.11 standard defines a distribution system as the infrastructure used to connect access points. To establish a distributed WLAN, two or more access points are configured with the same service set identifier (SSID). Access points configured with the same SSID make up a single logical network within a single Layer 2 broadcast domain, which means that they must all be able to communicate. The distribution system is the method used to connect them so they can do that.

The most common use of a wireless distribution system is to bridge a WLAN spanning two buildings. The simplest WDS consists of two access points configured to forward messages to each other, working in conjunction with an antenna that enables line-of-sight communication.

In a wired distribution system, APs typically connect through an Ethernet switch.

Wireless Application Protocol

The Wireless Application Protocol (WAP) is an open standard first developed by the members of the WAP Forum in 1997 for mobile applications in a wireless communications environment. The goal was to develop a technology that would operate over any kind of mobile wireless network, including GSM, CDMA and TDMA-based networks, and 3G technologies such as UMTS. The WAP forum, whose original members included Nokia, Ericsson, Unwired Planet and Motorola, is now part of the Open Mobile Alliance (OMA). The forum's membership now includes players such as Microsoft, IBM, Oracle and Intel.

WAP technology is primarily designed to enable mobile devices such as mobile phones and personal data assistants (PDAs) to access the World Wide Web and other Internet services such as email, information services and media content. It takes account of the need to optimize the use of the limited bandwidth available to such devices, minimize power requirements, and build in tolerance for the often unpredictable nature of wireless network availability. A micro-browser on the mobile device provides similar functionality to a web browser on a desktop computer, although its capabilities are limited due to the constraints imposed by the architecture of the mobile device (limited display space, processing power and memory). WAP content was originally created using the Wireless Markup Language (WML), an XML-based version of HTML optimized for wireless environment, and hosted on a standard web server. WAP also provided WML Script, the wireless equivalent of JavaScript.

The original WAP standard was version 1.0, released in 1998. It was closely followed by version 1.1 in 1999 and version1.2 in 2000. These early versions employed a protocol stack that was intended to completely replace the application and transport layer protocols used by the World Wide Web (primarily HTTP and TCP or UDP). The replacement protocols were similar in operation, and provided the same services as their standard counterparts, but were optimized for the constraints of a mobile wireless environment and designed to operate over any bearer network technology as opposed to only IP-based networks.

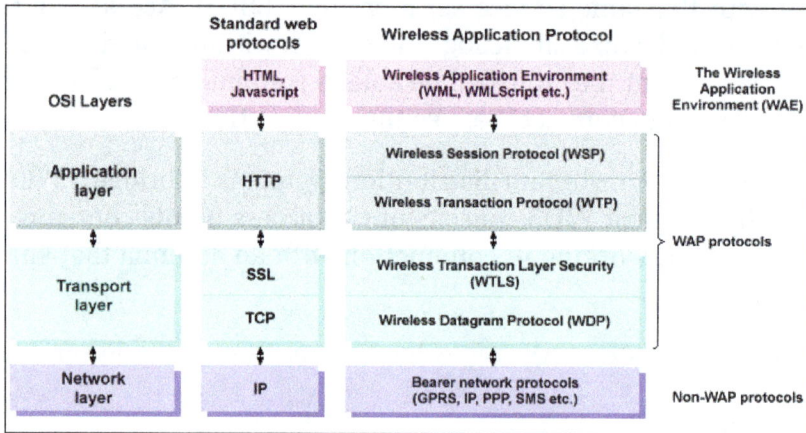

The WAP protocol stack.

With version 1.x of the wireless application protocol, a request generated by a mobile device for a WML document will normally be routed via a *WAP gateway*. The gateway provides the interface between the web server, to which it is connected via the Internet, and the mobile client, to which it connects via the client's wireless bearer network. It performs the required translation between the WAP protocols used by the client device and the HTTP and TCP/IP protocols used by the web server. It also compresses downstream data (i.e. WML and WML Script) for transmission across the wireless channel. The micro-browser on the client device decompresses and interprets the resulting byte-code and displays the results on the mobile device's display. The role of the WAP gateway is illustrated below.

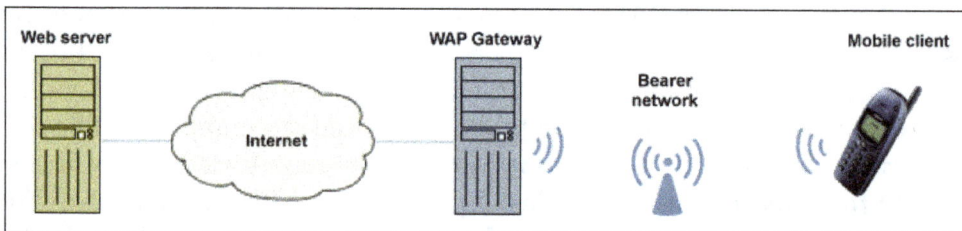

The WAP gateway carries out the necessary protocol translation and content encoding.

The WAP 1.x protocols enable communication to take place between the WAP gateway and the micro-browser. They are described below:

- Wireless Session Protocol (WSP) - With WTP, a complete replacement for HTTP, allowing the efficient exchange of data between mobile web applications.

- Wireless Transaction Protocol (WTP) - Provides transaction support (reliable request/response) to the datagram service provided by WDP.

- Wireless Transport Layer Security (WTLS) - An optional security layer based on the Secure Sockets Layer (SSL) protocol that provides a secure transport layer connection using encryption.

- Wireless Datagram Protocol (WDP) - A transport layer datagram service like UDP that sends and receives messages via any available bearer network, providing unreliable transport of data. The precise implementation of WDP is dependent on the network layer protocol used by the bearer network, although on IP-based networks it is effectively the same as UDP.

This first version of WAP was not an immediate success in either Europe or North America, although it fared rather better in Japan (possibly due to a more innovative approach to marketing and content provision). Authoring tools for WAP content were not widely available, and the level of support provided by service providers varied considerably. WAP version 2.0 went a long way towards remedying this situation, however, and WAP saw a revival of its fortunes between 2003 and 2004.

WAP Version 2.0

The radically re-engineered version of the Wireless Application Protocol was released in 2002 as version 2.0, and essentially replaces the original WAP protocol stack with standard Internet protocols. It does not therefore require a WAP gateway, but a proxy server is often used to optimise communication. A proxy can also improve access times and make more efficient use of network bandwidth by storing copies of frequently accessed resources. Current mobile devices have a much higher specification than the devices in use when WAP first appeared, enabling both HTTP and TCP to be used (although performance is often further enhanced through the use of the *wireless profile* version of both protocols).

The Wireless Markup Language (WML) has been replaced by *XHTML Mobile Profile* (XHTML MP), a cut-down version of XHTML. This, together with improved availability of web authoring tools that support XHTML MP, means that it has become easier to create web content that is suitable for mobile devices. The use of HTTP at the application layer has the additional advantage that multimedia content can be sent to a WAP device using the *Multimedia Message Service* (MMS), which has evolved from the *Simple Messaging Service* (SMS) used by millions of people every day to send text messages to each other. Another important concept introduced with WAP 2.0 is that of *push* – a mechanism that allows network servers to initiate the delivery of content to a WAP client device (as opposed to the normal *pull* model in which the server must wait for a request from the client device).

Personal Area Network

A personal area network (PAN) is the interconnection of information technology devices within the range of an individual person, typically within a range of 10 meters.

For example, a person traveling with a laptop, a personal digital assistant (PDA), and a portable printer could interconnect them without having to plug anything in, using some form of wireless technology. Typically, this kind of personal area network could also be interconnected without wires to the Internet or other networks.

Also see wireless personal area network (WPAN) which is virtually a synonym since almost any personal area network would need to function wirelessly. Conceptually, the difference between a PAN and a wireless LAN is that the former tends to be centered around one person while the latter is a local area network (LAN) that is connected without wires and serving multiple users.

In another usage, a personal area network (PAN) is a technology that could enable wearable computer devices to communicate with other nearby computers and exchange digital information using the electrical conductivity of the human body as a data network. For example, two people each wearing business card-size transmitters and receivers conceivably could exchange information by shaking hands.

The transference of data through intra-body contact, such as handshakes, is known as linkup. The human body's natural salinity makes it a good conductor of electricity. An electric field passes tiny currents, known as Pico amps, through the body when the two people shake hands. The handshake completes an electric circuit and each person's data, such as e-mail addresses and phone numbers, are transferred to the other person's laptop computer or a similar device. A person's clothing also could act as a mechanism for transferring this data.

Metropolitan Area Networks

A metropolitan area network (MAN) is a network with a size greater than LAN but smaller than a WAN. It normally comprises networked interconnections within a city that also offers a connection to the Internet.

The distinguishing features of MAN are:

- Network size generally ranges from 5 to 50 km. It may be as small as a group of buildings in a campus to as large as covering the whole city.

- Data rates are moderate to high.

- In general, a MAN is either owned by a user group or by a network provider who sells service to users, rather than a single organization as in LAN.

- It facilitates sharing of regional resources.

- They provide uplinks for connecting LANs to WANs and Internet.

Example of MAN

- Cable TV network.

- Telephone networks providing high-speed DSL lines.

- IEEE 802.16 or WiMAX, that provides high-speed broadband access with Internet connectivity to customer premises.

Global Area Network

A global area network (GAN) refers to a network composed of different interconnected networks that cover an unlimited geographical area. The term is loosely synonymous with Internet, which is considered a global area network.

Unlike local area networks (LAN) and wide area networks (WAN), GANs cover a large geographical area.

The Global Area Network is the web connecting various terminals and LANs together. This is used so that the data can be transferred from one point to another even if they are not directly connected. For this type of network, there is either a central server or all connected terminals act as a relay for the data to find its way to the end point. Uses of many wireless connection and satellite coverage.

The mobile GAN is the use of terminals over a wide geographical zone to act as relays for wireless connection.

Cellular Wireless Networks

Cellular network is an underlying technology for mobile phones, personal communication systems, wireless networking etc. The technology is developed for mobile radio telephone to replace high power transmitter/receiver systems. Cellular networks use lower power, shorter range and more transmitters for data transmission.

Features of Cellular Systems

Wireless Cellular Systems solves the problem of spectral congestion and increases user capacity. The features of cellular systems are as follows:

- Offer very high capacity in a limited spectrum.

- Reuse of radio channel in different cells.

- Enable a fixed number of channels to serve an arbitrarily large number of users by reusing the channel throughout the coverage region.

- Communication is always between mobile and base station (not directly between mobiles).

- Each cellular base station is allocated a group of radio channels within a small geographic area called a cell.

- Neighboring cells are assigned different channel groups.

- By limiting the coverage area to within the boundary of the cell, the channel groups may be reused to cover different cells.

- Keep interference levels within tolerable limits.

- Frequency reuse or frequency planning.

- Organization of Wireless Cellular Network.

Cellular network is organized into multiple low power transmitters each 100w or less.

Shape of Cells

The coverage area of cellular networks is divided into cells, each cell having its own antenna for transmitting the signals. Each cell has its own frequencies. Data communication in cellular networks is served by its base station transmitter, receiver and its control unit.

The shape of cells can be either square or hexagon:

Square

A square cell has four neighbors at distance d and four at distance Root 2d:

- Better if all adjacent antennas equidistant.

- Simplifies choosing and switching to new antenna.

Hexagon

A hexagon cell shape is highly recommended for its easy coverage and calculations. It offers the following advantages:

- Provides equidistant antennas.

- Distance from center to vertex equals length of side.

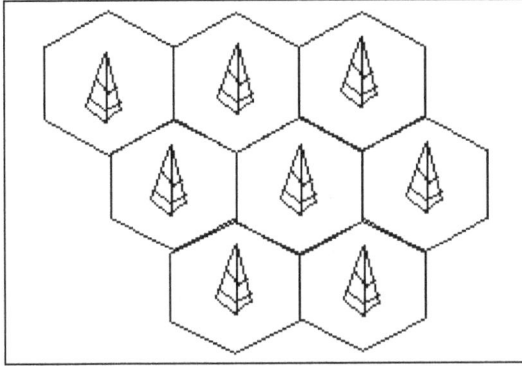

Frequency Reuse

Frequency reusing is the concept of using the same radio frequencies within a given area that are separated by considerable distance, with minimal interference, to establish communication.

Frequency reuse offers the following benefits:

- Allows communications within cell on a given frequency,
- Limits escaping power to adjacent cells,
- Allows re-use of frequencies in nearby cells,
- Uses same frequency for multiple conversations,
- 10 to 50 frequencies per cell.

For example, when N cells are using the same number of frequencies and K be the total number of frequencies used in systems. Then each cell frequency is calculated by using the formulae K/N.

Ad Hoc Wireless Network

"Ad hoc" means makeshift, or improvised, so a wireless ad hoc network (WANET) is a type of on-demand, impromptu device-to-device network. In ad hoc mode, you can set up a wireless connection directly to another computer or device without having to connect to a Wi-Fi access point or router.

Due to the nature of an ad hoc connection not needing an existing infrastructure to sustain the network, it's entirely decentralized and is considered a peer-to-peer network.

Instead of using a central managing device (like a router) where the network's data constantly flows in and out before and after reaching the child devices (like phones and computers), every node that makes up the ad hoc network forwards data evenly throughout the entire structure.

Wireless Ad Hoc Network Details

Following are some features, uses, benefits, and disadvantages of ad hoc networks:

- Expensive equipment isn't necessary to set up an on-the-fly, ad hoc network.

- There isn't a single point of failure in an ad hoc network.

- Ad hoc networks are useful when you need to share files or other data directly with another computer but don't have access to a Wi-Fi network.

- In emergency situations where a wireless network is suitable but there isn't an underlying network to use, wireless ad hoc networks deploy quickly and produce similar results.

- More than one laptop can be connected to the ad hoc network, as long as all the adapter cards are configured for ad hoc mode and connect to the same SSID. The computers need to be within 100 meters of each other.

- You can use an ad hoc wireless network to share your computer's internet connection with another computer.

- No central management hub where all devices can be controlled.

Types of Wireless Ad Hoc Networks

Wireless ad hoc networks are categorized into different classes. Here are a few examples:

- Mobile ad hoc network (MANET): An ad hoc network of mobile devices.

- Vehicular ad hoc network (VANET): Used for communication between vehicles. Intelligent VANETs use artificial intelligence and ad hoc technologies to communicate what should happen during accidents.

- Smartphone ad hoc network (SPAN): Wireless ad hoc network created on smartphonesvia existing technologies like Wi-Fi and Bluetooth.

- Wireless mesh network: A mesh network is an ad hoc network where the various nodes are in communication directly with each other to relay information throughout the total network.

- Army tactical MENT: Used in the army for "on-the-move" communication, a wireless tactical ad hoc network relies on range and instant operation to erect networks when needed.

- Wireless sensor network: Wireless sensors that collect everything from temperature and pressure readings to noise and humidity levels, can form an ad hoc network to deliver information to a home base without needing to connect directly to it.

- Disaster rescue ad hoc network: Ad hoc networks are important when disaster strikes and established communication hardware isn't functioning properly.

Ad Hoc Wireless Network Limitations

For file and printer sharing, all users need to be in the same workgroup, or if one computer is joined to a domain, the other users have to have accounts on that computer in order to access shared items.

Other limitations of ad hoc wireless networking include the lack of security and a slow data rate. Ad hoc mode offers minimal security; if an attacker comes within range of your ad hoc network, he or she won't have any trouble connecting.

References

- Wireless-communication-satellite, wireless-communication: tutorialspoint.com, Retrieved 13 May, 2019
- Principles-of-satellite-communications, principles-of-communication: tutorialspoint.com, Retrieved 4 August, 2019
- Wireless-distribution-system-WDS: searchmobilecomputing.techtarget.com, Retrieved 11 June, 2019
- Wireless-application-protocol, communication-technologies, telecommunications: technologyuk.net, Retrieved 9 February, 2019
- Personal-area-network: searchmobilecomputing.techtarget.com, Retrieved 11 April, 2019
- Metropolitan-Area-Networks-MAN: tutorialspoint.com, Retrieved 21 January, 2019
- Wireless-communication-cellular-networks, wireless-communication: tutorialspoint.com, Retrieved 3 July, 2019
- What-is-an-ad-hoc-wireless-network-2377409: lifewire.com, Retrieved 25 March, 2019

Wireless Sensor Networks

The group of spatially separated sensors which monitor and record the physical conditions of the environment and then organize the data in a central location is known as wireless sensor network. They can be used for health care monitoring and air pollution monitoring. The diverse aspects and applications of wireless sensor networks have been thoroughly discussed in this chapter.

Wireless Sensor Networks (WSNs) can be defined as a self-configured and infrastructure-less wireless networks to monitor physical or environmental conditions, such as temperature, sound, vibration, pressure, motion or pollutants and to cooperatively pass their data through the network to a main location or sink where the data can be observed and analyzed. A sink or base station acts like an interface between users and the network. One can retrieve required information from the network by injecting queries and gathering results from the sink. Typically a wireless sensor network contains hundreds of thousands of sensor nodes. The sensor nodes can communicate among themselves using radio signals. A wireless sensor node is equipped with sensing and computing devices, radio transceivers and power components. The individual nodes in a wireless sensor network (WSN) are inherently resource constrained: they have limited processing speed, storage capacity, and communication bandwidth. After the sensor nodes are deployed, they are responsible for self-organizing an appropriate network infrastructure often with multi-hop communication with them. Then the onboard sensors start collecting information of interest. Wireless sensor devices also respond to queries sent from a "control site" to perform specific instructions or provide sensing samples. The working mode of the sensor nodes may be either continuous or event driven. Global Positioning System (GPS) and local positioning algorithms can be used to obtain location and positioning information. Wireless sensor devices can be equipped with actuators to "act" upon certain conditions. These networks are sometimes more specifically referred as Wireless Sensor and Actuator Networks.

Bluetooth

Bluetooth is a telecommunications industry specification that describes how mobile devices, computers and other devices can easily communicate with each other using a short-range wireless connection.

Early Bluetooth versions allowed users of cellular phones, pagers and personal digital assistants to buy a three-in-one phone that could double as a portable phone at home or in the office, get quickly synchronized with information in a desktop or notebook computer, initiate the sending or receiving of a fax, initiate a printout, and in general, have all mobile and fixed computer devices be totally coordinated over a short distance.

More recent Bluetooth versions make it possible for a user to place hands-free phone calls through a mobile phone or connect wireless headphones to a smartphone's music playlist, for example. Bluetooth technology can simplify tasks that previously involved copious wires strewn among peripheral devices. For instance, with a Bluetooth-enabled printer, one can connect wirelessly with a desktop, laptop or mobile device and print out documents. It is also possible to sync a wireless keyboard with a tablet-style device, such as a DVD player with a television.

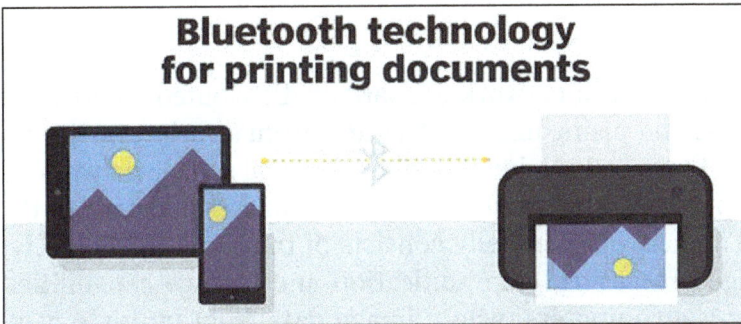

Bluetooth technology for printing documents

Additionally, mobile operating systems allow users to stream media, such as movies, television shows and music, to compatible TVs, speakers and media players via Bluetooth. With an eye toward the future of Bluetooth, companies such as LG are manufacturing televisions with built-in Bluetooth technology that can display 3D images users view through special "active-shutter" glasses. Though this technology is in its formative stages, it's gotten an enthusiastic reception from gamers.

Laptop or desktop computers without built-in Bluetooth can gain those capabilities through an inexpensive USB dongle. The one caveat here is Bluetooth technology typically uses considerable battery power, so it's suggested that it be monitored closely by the user to prevent a device's battery from running down.

Working of Bluetooth

Bluetooth technology requires that a low-cost transceiver chip be included in each device. The transceiver transmits and receives in a previously unused frequency band of 2.45 GHz that is available globally - with some variation of bandwidth in different countries. In addition to data, up to three voice channels are available. Each device has a unique 48-bit address from the IEEE 802 standard. Bluetooth connections can be point to point or multipoint.

The maximum Bluetooth range is 10 meters. Data can be exchanged at a rate of 1 megabit per second - up to 2 Mbps in the second generation of the technology. A frequency hop scheme allows devices to communicate even in areas with a great deal of electromagnetic interference. Built-in encryption and verification is provided.

Visual Sensor Network

A visual sensor network is a network of spatially distributed smart camera devices capable of processing and fusing images of a scene from a variety of viewpoints into some form more useful than the individual images. A visual sensor network may be a type of wireless sensor network, and much of the theory and application of the latter applies to the former. The network generally consists of the cameras themselves, which have some local image processing, communication and storage capabilities, and possibly one or more central computers, where image data from multiple cameras is further processed and fused (this processing may, however, simply take place in a distributed fashion across the cameras and their local controllers). Visual sensor networks also provide some high-level services to the user so that the large amount of data can be distilled into information of interest using specific queries.

The primary difference between visual sensor networks and other types of sensor networks is the nature and volume of information the individual sensors acquire: unlike most sensors, cameras are directional in their field of view, and they capture a large amount of visual information which may be partially processed independently of data from other cameras in the network. Alternatively, one may say that while most sensors measure some value such as temperature or pressure, visual sensors measure *patterns*. In light of this, communication in visual sensor networks differs substantially from traditional sensor networks.

Applications

Visual sensor networks are most useful in applications involving area surveillance, tracking, and environmental monitoring. Of particular use in surveillance applications is the ability to perform a dense 3D reconstruction of a scene and storing data over a period of time, so that operators can view events as they unfold over any period of time (including the current moment) from any arbitrary viewpoint in the covered area, even allowing them to "fly" around the scene in real time. High-level analysis using object recognition and other techniques can intelligently track objects (such as people or cars) through a scene, and even determine what they are doing so that certain activities could be automatically brought to the operator's attention. Another possibility is the use of visual sensor networks in telecommunications, where the network would automatically select the "best" view (perhaps even an arbitrarily generated one) of a live event.

Sensor Grid

Sensor Grid is a combination of a physical grid with an embedded detection sensor. The detection technology can be either opti-grid (fiber optic) or cast wire, both of which are optimized to detect cutting or bending of the structure.

The heavy-gauge steel grid is very robust and will sustain heavy water flow. It remains operational even when completely submerged under water for many years.

Application

Each Sensor Grid segment is carefully tailored to the specific size and nature of the opening. Based on the application (water pass, window grid, etc.), each segment can be either fixed or movable on a hinge or rail. For example, a sliding rail for a gate can be designed for water passage, to automatically raise the grid when water levels reach a certain height. Sensor Grid typically complements a larger PIDS application. The communication system should be integrated with other sensors, feeding a synchronized data stream into the Security Management System (SMS).

Sensor Grid is an enhanced conventional physical barrier offering state-of-the-art protection. Any attempt to cut or remove part of the grid or the grid itself, is immediately detected.

Sensor Grid comes in two versions:

- CAST - Sensor Grid uses an embedded electro-mechanical sensor, which detects cutting and bending of the steel grid. It requires no power and uses a standard dry contact output.

- OPTIGRID - Sensor Grid uses an electro-optical sensor, which detects cutting and bending of the steel grid. It is connected to a communication processor and communicates through a standard dry contact output or via a long range RS-422 serial output.

Sensor Grid grids are built to order to suit specific customer requirements.

In the case of Sensor Grid installations where occasional access is required, Sensor Grid grids can be installed on rails and integrated with winches to enable lifting.

Sensor Grid can be integrated with any SMS system that can accept dry contact inputs or the CCC RS-422 protocol, including third-party systems.

Wireless Sensor Networks for Environmental Monitoring

Environmental monitoring applications can be broadly categorized into indoor and outdoor monitoring. Indoor monitoring applications typically include buildings and offices monitoring. These applications involve sensing temperature, light, humidity, and air quality. Other important indoor applications may include fire and civil structures deformations detection. Outdoor monitoring applications include chemical hazardous detection, habitat monitoring, traffic monitoring, earthquake detection, volcano eruption, flooding detection and weather forecasting. Sensor nodes also have found their applicability in agriculture. Soil moisture and temperature monitoring is one of the most important application of WSNs in agriculture. Only outdoor environmental monitoring will be considered in this work.

The WSN architecture consists of a Reduced Instruction Set Computer (RISC) microcontroller with a small program and data memory. An external flash memory can be used to provide secondary storage. Two approaches have been adopted for the design of sensing equipment. The first approach uses a sensing board that can be attached to the main microcontroller board through an expansion bus. Usually, more than one can be attached. Other boards only have I/O (input-output) connectors and can be used to connect custom sensor to the main board. In the second approach, the main board also includes the sensing devices. The sensing devices are soldered or can be mounted if needed.

Here, we have used the Atmega 128 processor, with 128 KB flash memory, 4 KB RAM, and a stream-based Nordic nRF903 radio transceiver of 433 MHz, providing 72 Kb/s channel and a range of 500m using quarter-wave antenna. The board includes power supply, solar charging circuit, and sensing for on-board temperature, battery voltage, and charging current. Battery, solar cell, serial, and analog and digital transducers could be connected using just a screwdriver- no expansion board was required. The BMAC protocol is used here and Tiny OS operating system is used which is the most used open source and freeware WSN operating systems. In addition the sensor nodes are provided with the cameras attached to, it to capture the images of the key locations. This deployment had raised new challenges to routing and network topology.

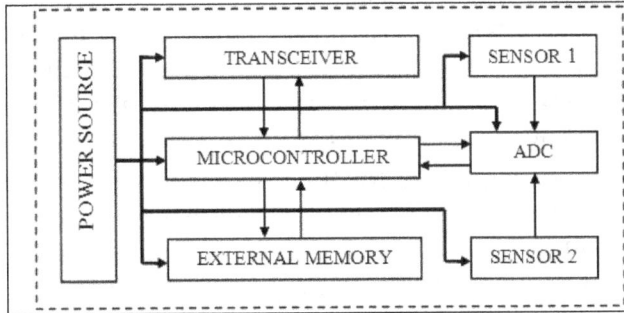

Wireless Sensor Network Architecture.

Applications

We discuss the wireless sensor network applications that have been introduced over past six years, the technical challenges they faced and what we have learned from them. These applications follow a common theme: understanding the natural and agricultural environments. This includes monitoring of farms and rain forests, cattle monitoring, agricultural monitoring, water quality monitoring, volcanic eruptions and earth-quake monitoring etc. Our design choices and technologies were made in accordance to meet the different challenges presented by each application.

Cattle Monitoring

We organized a network for research at a farm over 500 km from our lab. The first phase of our project covers the positions of cattle over time, and soil moisture at different points in the farm. Soil moisture measurement shows how quickly the pasture will grow and hence how many animals required per unit area. The information from static and mobile nodes are conveyed to the base station and then over the internet to a remote server. The cameras that attached to the sensor nodes periodically capture the images of key locations at the farm. The Fleck nodes were built into collars that were worn by the cattle and these nodes was specialized for animal tracking applications.

Ground Water Quality Monitoring

For this purpose, we developed a relatively small network, located 2000 km from our lab. Its purpose was to monitor the salinity, water level, and water extraction rate at a number of bores. This is a coastal region where over extraction of water leads to saltwater intrusion into the aquifer.

Lake Water Quality Monitoring

The purpose of our project was to measure vertical temperature profile at multiple points on a large water storage that provides most of the drinking water for the city. The data, from a string of temperature transducers at depths from 1 to 6 m at 1-m intervals, provide information about water mixing within the lake which can be used to predict the development of algal blooms. Low-power wireless communications over water proved to be a challenge due to multipath as radio waves reflected from the water surface destructively interfere with waves traveling directly. Interfacing a robotic boat to the static sensor nodes was another challenge. The network comprises floating sensor nodes and a custom expansion board for the one-wire temperature transducer string. The node is mounted on an anchored float, along with a solar cell and a high-gain whip antenna. The most novel element in this network is a solar-powered robotic boat. Navigation is by GPS and depth sounder, and a scanning laser range-finder mounted high and looking forward detects obstacles.

Rainforest Monitoring

The major initiative is to provide reliable, long term monitoring of rainforest ecosystems and also monitoring the restoration of biodiversity. The first phase of the project concerned with the long-term, low-power WSNs in rainforest environment. As the solar energy was very limited, we had to first quantify the performance of current WSN technology, in order to develop the network and energy management protocols required for robust and reliable performance of long-term rainforest networks. We implemented a low-power MAC protocol, to help meet the power budget. The nodes used the Fleck boards and environmental housing as shown in figure custom expansion board was built to interface to the many transducers: wind speed and direction, leaf wetness, soil moisture, temperature, and relative humidity.

Wireless sensor network.

Volcano Monitoring

A different type of extreme environment is targeted in the volcano monitoring system. In this application WSNs are equipped with low-frequency acoustic sensors to monitor volcanic activity. While traditional systems involved local storage of data, which thus required a manual collection of the sampled data for further processing, the WSN-based system allows real-time monitoring of the activity over wireless links. In addition to the continuous monitoring of the volcanic activity, the researchers implemented an event-detection mechanism to reduce the amount of data which had to be communicated and processed.

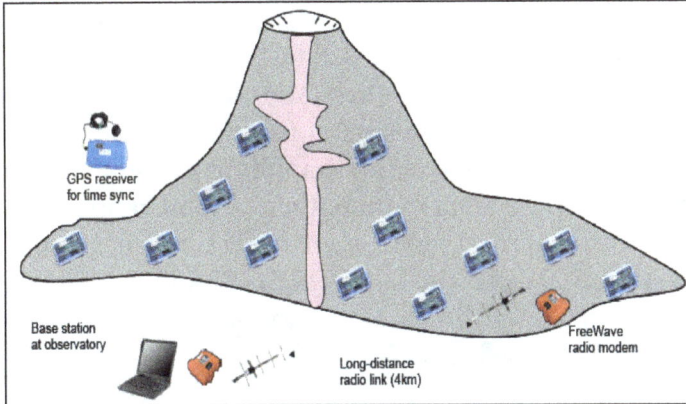

Wireless Sensor Networks for Volcano Monitoring.

Challenges for Environmental Wireless Sensor Networks

Although, extensive efforts have been done, there are still some challenges that need to be addressed. Power-management is essential for long-term operation, especially when it is needed to monitoring remote and hostile environments. In terms of scalability, a wireless sensor network can accommodate thousands of nodes. Systems installed on isolated locations cannot be visited regularly, so a remote access standard protocol is necessary to operate, to manage, to reprogramming and to configure the WSN, regardless of the manufacturer. The WSN need to become easier to install, maintain and understand. The storage capacity and redundancy can be increased by adding nodes to the system. Increasing the storage nodes and configuring them to capture overlapping areas of the sensor nodes ensures that there are multiple copies of the data, thus providing redundancy in case some of the storage nodes fail. Reducing the size is essential for many applications. Battery size and radio power requirements play an important role in size reduction. Producing cheaper, reliable, and disposable sensor platforms is also a big challenge.

Health Care Monitoring

Population around the world is ageing this is due to number of factors such as decreased fertility rates and increased expectancy of life, mainly linked to the increased birth control and movements of migration. This trend is more evident and continuously growing in developed countries, because the number of elderly people is already large. Aging has impact on the social and economic foundations of communities, thus; governments need to employ more funding for taking care of the elderly people and fewer hands are available cover their needs.

For the health care systems, this is considered as a big challenge, these health care systems need to deal with the health issues of the elderly population. In order to develop effective health systems, they must be able to deal with the so-called "four giants" of geriatrics (Isaacs) as: instability, immobility, intellectual impairment and incontinence. Moreover, chronic diseases are widely spread in elderly, such as Alzheimer's, chronic respiratory diseases, cardiovascular diseases and diabetes. Thus, mobile application widely used in our daily life. On of reasons of using mobile application in education, government services and health care is the usability, ease of use and usefulness of using the application at anywhere and anytime. Using mobile application to track elderly people is one of the benefits of such services. On other hand, accessing information and monitoring the up-to-date issues is the best benefit of mobile application. Using mobile application in health care would help doctors follow up with patient and track their status without the hassle of traveling to hospital.

According to the World Health Organization, elderly chronic diseases are the main cause of death in the world, accounting 60% of all death. In Europe this proportion is higher, where these diseases are estimated to account for 77% of the disease burden and an 86% of the total deaths in the area. Thence, it is essential to develop effective health systems that are able to keep the chronic diseases under control.

The Active Ageing, which is the process of improving opportunities for elderly's health, security and participation which aims to improve quality of life as individuals age, is portion of the process of handling these chronic diseases, to make effort to keep the elderly healthy and participating in the social live. Recently; artificial intelligence algorithms were widely used for solving very complicated problems, such as patterns recognition and information retrieval, image segmentations and river flow forecasting. Many researchers employed communication technologies and information and artificial intelligence algorithms for covering real health needs of patients.

The proposed EHC aims to develop an integrated and multidisciplinary method to employ communication technologies and information for covering real health needs of elderly people. The system will help the elderly to prevent or keep under control their chronic diseases as well as make the physician's aware of the chronic diseases of the elderly and help them when necessary. The work will be useful for the professionals giving them ways for treatments' prescription (medical treatments) for the elderly patients and for a following up of the patients. Reduce the chance of life taken away. This work also profits the health control products, which are incorporated within the systems to enable treatments' follow up.

A pervasive monitoring system to transfer the patients physical signs to real time remote medical applications was proposed in. The system had two components: data acquisition section and data transmission section. The health parameters (ECG, blood pressure, heart rate, SpO_2, blood fat, pulse rate and blood glucose) in addition to other indicator (location of the patient) are planned to be sampled continuously at different rates. Four modes of data transmission are used medical analysis needs, taking risk, computing resources into consideration and demands for communication. Finally, the authors implemented a sample prototype to provide a system overview.

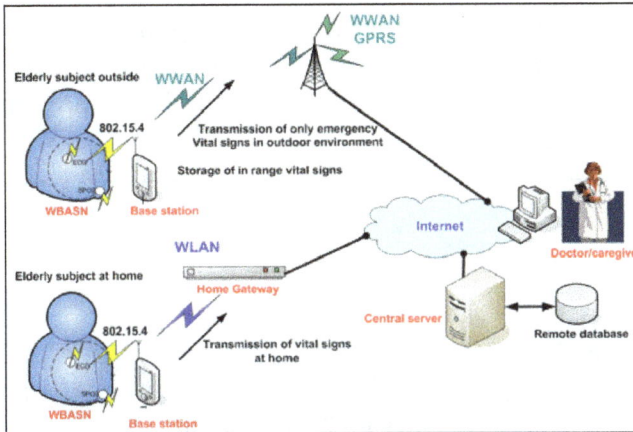

The monitoring system architecture.

Ronnie et al., in offered a framework for u-Healthcare system using Radio Frequency Identification (RFID) and Wireless Medical Sensor Networks (WMSN). The system of monitors the medical status of the patient by using body sensor with the RFID and transmits the data by the wireless to the nearest local workstation (WMSN gateway) then transmits it to the central server. Proper medical services are administered locally in the workstation depending on queries with the central database that has the patient's information. In their work, patients are alerted when there is an emergency case and receive alert message with their smart phones. The medical staff on the workstation will also receive a messages showing that the patient's health needs attention. Based on the patient's health status a medical service will be applied or prescribed to the patient.

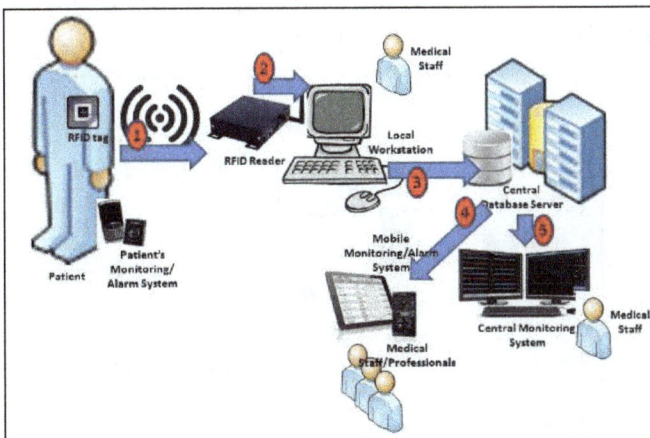

The U-healthcare system design using WMSN and RFID.

Al-Aubidy et al., in designed and implemented a real time healthcare monitoring system. The system scans, calculates, monitors and communicates with medical center's using a group of sensors connected to microcontroller which has a tool for wireless communication to transmit the real time health information from the patient to the medical center which helps to detect any abnormal medical condition.

Elderly health monitoring system using smart home gateway was proposed in. The system has provided continuous and long-term monitoring for the elderly. In consideration of the mass data generated in the monitoring process, an ECG compression algorithm was built and tested. The results show that it is possible to apply the algorithm to the real-time monitoring system.

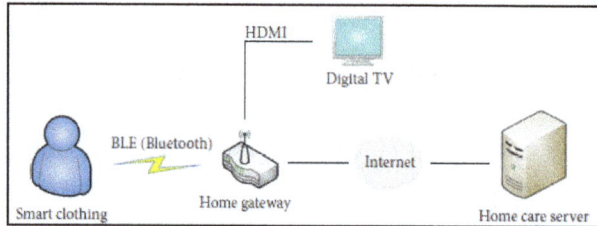

The Architecture health monitoring system.

Gogate and Bakal in presented a healthcare model using WSN to remove limitations of wired system and to utilize wireless technology efficiently. The system continuously screens the patient's body temperature and heart rate and when detecting any abnormal parameter, sending email alert or sms to the physician, nurse and close relative of the patient for emergency handling.

Proposed EHC Healthcare Expert System Design

The EHC expert system can be divided into three sections; the smart bracelets with embed medical sensors and it uses wireless technology to communicate with DB server, and a LAN that connect the medical center unit which is consist of DB server and nurses' computers, the third section is the physicians' PDAs connected to the DB server through internet infrastructure communications.

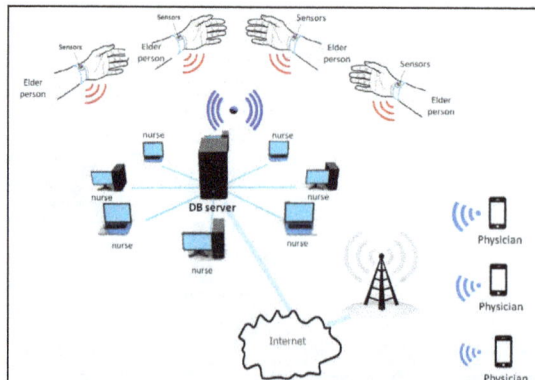

The proposed healthcare expert system architecture design.

Each patient smart bracelet is considered as a node of a Wireless Sensors Networks (WSN) with Gateway node at DB server. Each nurse computer connected to the DB server node through LAN connection. The DB server node receives the health indicator measurements from the bracelets; it stores the measurements, and sends these measurements to nurses' computers. In the case of critical measurements recorded, flags sent by the DB server to the physician's PDAs.

The bracelet unit is considered as wireless sensor node, and it is equipment with microcontrollers and embed health indicator sensors and as shown in figure. The EHC designed with the use of low power and micro medical indicator sensors. They are four medical indicator sensors to measure the body temperature (BT), blood pressure (BP), and Heart beat rate (HR).

The physician can have checked the elder health syndrome by using his PDA or his PC through internet connection to the DB server.

The DB server employs a healthcare software system to analyze the received sensed indicator, that software is equipment with an expert system for health care domain, it functionality can be described as follows:

- Receiving the indicator values from the bracelet sensors.

- Analyze the measurements to define any abnormal or vital health condition.

- Illustrate the received data and analyzed information of the elders' health condition to the nurses' computers.

- Determine if any necessary condition must be reported to the physicians' PDAs using the expert system.

- Get any recommendation by the physician and added to the DB and also update the expert system knowledge base if needed.

The database system contains table of the elder to show the personal information of each elder with the ability of find a specific person by his name and his ID value. When a elder is selected, his/her information can be modified, and if there is any abnormal measurements indicators, that will appear and stored as a note. Besides that, it provides the date and time of last update of his biomedical data. Medical conditions query shows the data for the specified patient, such as all normal and abnormal readings stored in the database.

The medical staff on their workstations will receive an alert message indicating that the patient status if it needs attention. As well as, Physician will be alerted in case of emergency by sending a message to their PDAs. A medical service depending on the status of the patient will be prescribed or applied to the patient.

When the physician login to his application interface, he will be directly being able to see the data received by the system for every individual patient along with their current and previous medical reports in the database system with their statistical charts. The physician can view the patient medical history on the interface application on his PDA. So, based on the previous medical records and current measurements the physician can then prescribe the medication in acknowledgement to the patients' data.

System Discussion

By getting the correct information about the elderly health at the right time, the proposed EHC system will help the nurses and physician through following advantages:

- Assists the physician in tracking and monitoring the elderly health from a distance.

- Gives alerts about changes in the elderly health and reduce the complications.

- Monitors the condition of the elderly periodically.

- Contribute to saving the lives of elderly.

- Help the nurse to ask for help through the application by the alert that appears to the physician's when any sudden change in (pressure - heartbeat - body temperature - sugar) happens.

The prototype of the proposed EHC system has been verified and tested to be sure that all of the software and hardware components are working accurately. The smart bracelet with embedded medical sensors has a wireless technology to connect the elderly side with DB server, as well as the physicians' PDAs are connected to the DB server through internet infrastructure communications. The programmed database contains the patient's details, medical conditions, medical history, medications and patient's health status evaluation. Moreover; the database contains the real data received from smart bracelets including the BT, BP and HR. the user interface was designed to help the nurses and physicians to monitor and track the elderly's real-time health parameters. The main limitation of this proposed EHC is the smart bracelet with set of integrated medical sensors, due to the limited fund for this project the smart bracelet in this stage is not fully functionally work. Therefore; in this stage we used a nurse to replace the smart bracelet roles. The future work is to design and implement the smart bracelet with set of integrated medical sensors for elderly healthcare monitoring system to transfer real-time medical information between the elderly and the medical center.

Air Pollution Monitoring System

Sensor networks are dense wireless networks of small, low-cost sensors, which collect and disseminate environmental data. Wireless sensor networks facilitate monitoring and controlling of physical environments from remote locations with better accuracy. They have applications in a variety of fields such as environmental monitoring, indoor climate control, surveillance, structural monitoring, medical diagnostics, disaster management, and emergency response, ambient air monitoring and gathering sensing information in inhospitable locations. Sensor nodes have various energy and computational constraints because of their inexpensive nature and ad-hoc method of deployment. Considerable research has been focused at overcoming these deficiencies through more energy efficient routing, localization algorithms and system design.

Indeed, with the increasing number of vehicles on our roads and rapid urbanization air pollution has considerably increased in the last decades in Mauritius. For the past thirty years the economic development of Mauritius has been based on industrial activities and the tourism industry. Hence, there has been the growth of industries and infrastructure works over the island. Industrial combustion processes and stone crushing plants had contributed to the deterioration of the quality of the air.

Mauritius has led to a major increase in the number of vehicles on the roads, creating additional air pollution problem with smoke emission and other pollutants.

Air pollution monitoring is considered as a very complex task but nevertheless it is very important. Traditionally data loggers were used to collect data periodically and this was very time consuming and quite expensive. The use of WSN can make air pollution monitoring less complex and more instantaneous readings can be obtained. Currently, the Air Monitoring Unit in Mauritius lacks resources and makes use of bulky instruments. This reduces the flexibility of the system and makes it difficult to ensure proper control and monitoring. WAPMS will try to enhance this situation by being more flexible and timely. Moreover, accurate data with indexing capabilities will be able to obtain with WAPMS. The main requirements identified for WAMPS are as follows:

- Develop an architecture to define nodes and their interaction;
- Collect air pollution readings from a region of interest;
- Collaboration among thousands of nodes to collect readings and transmit them to a gateway, all the while minimizing the amount of duplicates and invalid values;
- Use of appropriate data aggregation to reduce the power consumption during transmission of large amount of data between the thousands of nodes;
- Visualization of collected data from the WSN using statistical and user-friendly methods such as tables and line graphs;
- Provision of an index to categorize the various levels of air pollution, with associated colors to meaningfully represent the seriousness of air pollution;
- Generation of reports on a daily or monthly basis as well as real-time notifications during serious states of air pollution for use by appropriate authorities.

At present, our scientific understanding of air pollution is not sufficient to be able to accurately predict air quality at all times throughout the country. This is where monitoring can be used to fill the gap in understanding. Monitoring provides raw measurements of air pollutant concentrations, which can then be analysed and interpreted. This information can then be applied in many ways. Analysis of monitoring data allows us to assess how bad air pollution is from day to day, which areas are worse than others and whether levels are rising or falling. We can see how pollutants interact with each other and how they relate to traffic levels or industrial activity. By analysing the relationship between meteorology and air quality, we can predict which weather conditions will give rise to pollution episodes.

Wireless Sensor Network (WSN) is an active field of research due to its emerging importance in many applications including environment and habitat monitoring, health care applications, traffic control and military network systems . With the recent breakthrough of Micro-Electro Mechanical Systems (MEMS) technology whereby sensors are becoming smaller and more versatile, WSN promises many new application areas in the near future. Typical applications of WSNs include monitoring, tracking and controlling. Some of the specific applications are habitat monitoring, object tracking, nuclear reactor controlling, fire detection, traffic monitoring, etc.

Initial development into WSN was mainly motivated by military applications. However, WSNs are now used in many civilian application areas for commercial and industrial use, including environment and habitat monitoring, healthcare applications, home automation, nuclear reactor controlling, fire detection and traffic control. This transition from the use of WSN solely in military applications has been motivated due to the nature of WSNs which can be deployed in wilderness areas, where they would remain for many years, to monitor some environmental variables, without the need to recharge/replace their power supplies. Such characteristics help to overcome the difficulties and high costs involved in monitoring data using wired sensors.

Fire and Flood Detection

Large number of environmental applications makes use of WSNs. Sensor networks are deployed in forest to detect the origin of forest fires. Weather sensors are used in flood detection system to detect, predict and hence prevent floods. Sensor nodes are deployed in the environment for monitoring biodiversity.

The Forest-Fires Surveillance System (FFSS) was developed to prevent forest fires in the South Korean Mountains and to have an early fire-alarm in real time. The system senses environment state such as temperature, humidity, smoke and determines forest-fires risk-level by formula. Early detection of heat is possible and this allows for the provision of an early alarm in real time when the forest-fire occurs, alerting people to extinguish forest-fires before it grows. Therefore, it saves the economic loss and environment damage. Similarly, a typical application of WSN for flood detection and prevention is the ALERT system deployed in the US. Rainfalls, water level and weather sensors are used in this system to detect, predict and hence prevent floods. These sensors supply information to a centralized database system in a pre-defined way.

Biocomplexity Mapping and Precision Agriculture

Wireless sensor networks can be used to control the environment which involves monitoring air, soil and water. Sensors are deployed throughout the field and these sensors form a network that communicate with each other to finally reach some processing centre which analyse the data sent and then accordingly adjust the environment conditions (e.g., if the soil is too dry, the processing centre send signals which actuators recognise accordingly and thus can start the sprinkling system). Biocomplexity mapping system helps to control the external environment. Sensors are used to observe spatial complexity of dominant plant species. An example is the surveillance of the marine ground floor where an understanding of its erosion processes is important for the construction of offshore wind farms.

Precision agriculture is an emerging WSN application area to monitor and control the amount of pesticides present in drinking water, monitor the level of soil erosion and the level of air pollution. Precision agriculture encompasses different aspects such as monitoring soil, crop and climate in a field. Huge amount of sensor data from large-scale agricultural fields are frequently generated in such an application.

Habitat Monitoring

Concerns associated with the impacts of human presence in monitoring plants and animals in field conditions have to a large extent been overcome by WSNs. Sensors can now be deployed prior to the onset of the breeding season and while plants are dormant or the ground is frozen as well as on small islets where it is unsafe or unwise to repeatedly attempt field studies. Such deployment represents a substantially more economical method for conducting studies than traditional personnel-rich methods where substantial proportion of logistics and infrastructure must be devoted to the maintenance of field studies, often at some discomfort and occasionally at some real risk.

Perhaps the best known application demonstrator for WSN in this domain is the Great Duck Island project at Berkley. Sensors monitored the microclimates in and around nesting burrows used by the Leach's Storm Petrel in a non-intrusive and non-disruptive manner. Motes were deployed on the island, with each of them having a microcontroller, a low-power radio, memory, and batteries. Readings such as Infrared levels, humidity, rainfall and temperature were monitored on a constant basis to better understand the movements of the petrels. Motes periodically sampled and relayed their sensor readings to computer base stations on the island which in turn fed into a satellite link that allows researchers to access real-time environmental data over the Internet.

Researchers at University of Florida and University of Missouri, Colombia are studying the role of wildlife in maintaining diversity, tracking invasive species and the spread of emerging diseases by obtaining unobtrusive visual information. They are using Deer Net which is a WSN-based system for analysing wildlife behavior by tracking deer's actions. The overall goal is to develop a long-lived and unobtrusive wildlife video monitoring system capable of real-time video streaming. The captured video will be transmitted over to a remote monitoring center for real-time viewing and camera control. Advanced scene classification and object recognition algorithms together with fusion of data from other sensors like GPS and motion can be applied to remove essential visual information from the captured video. Then, statistical models about animals' food selection, activity patterns and close interactions can be made consequently.

Recursive Converging Quartiles Data Aggregation Algorithm

Most wireless sensor networks involve the collection of high amounts of data. For this reason, during last year's considerable research effort has been devoted to data fusion and aggregation algorithms. In general, if we consider the problem to route data packets, representing measurements collected by sensors, to a single managing entity, i.e., a network sink, it is often efficient to exploit the correlation among similar data collected by the sensors in order to decrease overhead. At this point, however, a trade-off arises between the amount of transmitted data in the aggregated flows and their reliability. Data aggregation is a technique which tries to alleviate the localized congestion problem. It attempts to collect useful information from the sensors surrounding the event. It then transmits only the useful information to the end point thereby reducing congestion and its associated problems. We have developed a new data aggregation algorithm for WAPMS named Recursive Converging Quartiles (RCQ). The algorithm includes two basic operations namely duplicate elimination and data fusion.

Duplicate Elimination Technique

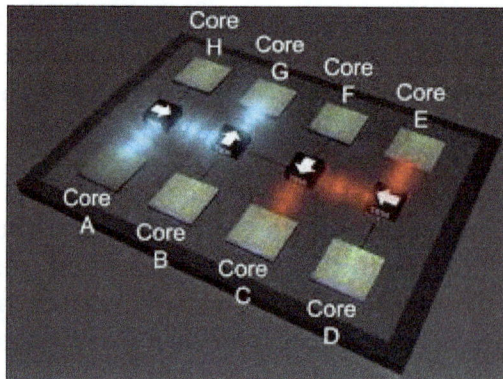

Multichip routing of data.

In WAPMS a packet consists of two parts: the data, which is the reading collected by the source node, and an id, which identifies the node uniquely in the network such as a network address. The cluster head collects readings from every node and stores them in a list. After collection, it goes through each item in the list and check for the occurrence of packets with the same id, thereby detecting the presence of duplicate packets. It then keeps only one instance of them. figure illustrate our proposed duplication elimination technique.

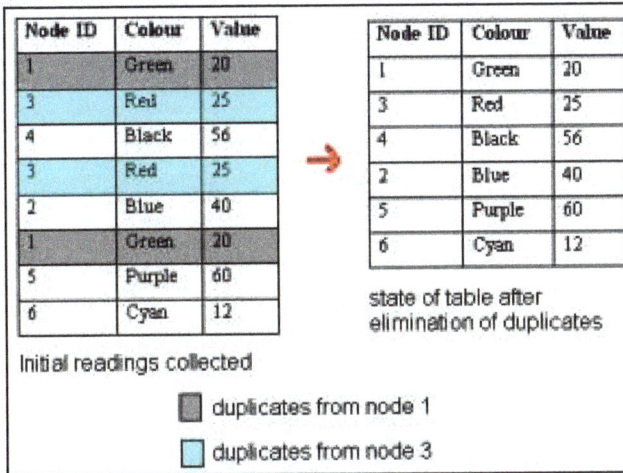

Node ID	Colour	Value
1	Green	20
3	Red	25
4	Black	56
3	Red	25
2	Blue	40
1	Green	20
5	Purple	60
6	Cyan	12

Initial readings collected

Node ID	Colour	Value
1	Green	20
3	Red	25
4	Black	56
2	Blue	40
5	Purple	60
6	Cyan	12

state of table after elimination of duplicates

◼ duplicates from node 1

☐ duplicates from node 3

Illustration of duplicate elimination technique.

Proposed Data Fusion Technique

There are several statistical methods to summarize a list of data. We have considered the use of the three quartiles - lower, median and upper. We have considered the use of quartiles since they are unaffected by extreme values; this is required in our system whereby extreme and invalid values can sometimes be transmitted to the cluster head and these should not influence the data fusion mechanism. Moreover, quartiles reduce the amount of data to only three values while still reflecting the original data in an accurate way. The novel data fusion algorithm works as follows:

- The list is partitioned into several smaller groups:
  - We consider the length of the list.
  - We find its multiples in the form (x_1, y_1), (x_2, y_2)...
  - E.g., length = 200, multiples = (1, 200), (2, 100), (4, 50), (5, 40), (10, 20), (20, 10), (24, 5).
  - We choose the pair which will give the highest number of groups (Maximise x) and the minimum number of elements per group, while keeping it above a threshold (Minimise y, y > threshold value) E.g., length = 50, multiples = (1, 50), (2, 25), (5, 10), (10, 5), threshold = 5, optimal pair = (10, 5).
- We calculate the quartiles for each of the smaller lists.
- Merge the resulting quartiles for the sub lists into one list.

- Repeat the whole process until the eventual number of groups, in which the list can be broken, becomes one and the final list obtained has only three values.

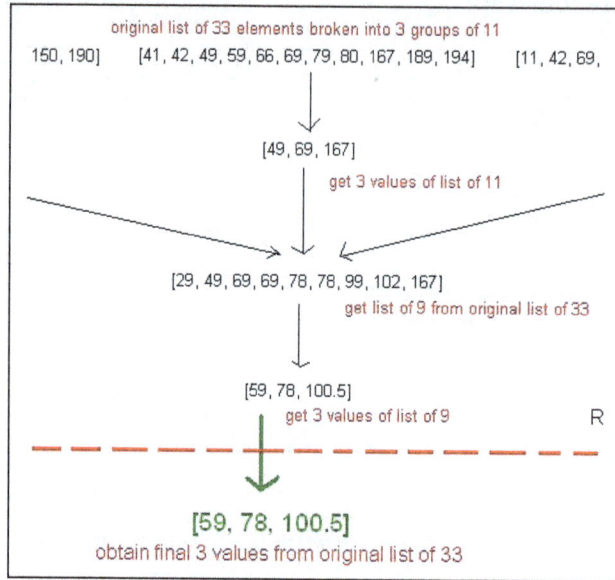

Using RCQ to aggregate a list of 33 values to only 3 values.

WAPMS: The Proposed Air Pollution Monitoring System

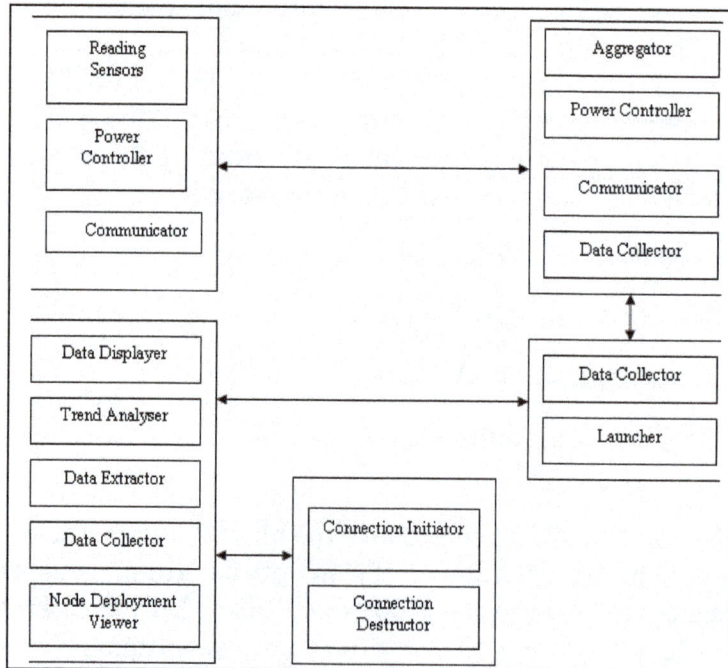

Architecture diagram of WAMPS.

The proposed wireless sensor network air pollution monitoring system (WAPMS) comprises of an array of sensor nodes and a communications system which allows the data to reach a server. The sensor nodes gather data autonomously and the data network is used to pass data to one or more base stations, which forward it to a sensor network server. The system send commands to the nodes in order to fetch the data, and also allows the nodes to send data out autonomously. Figure shows the architecture diagram of WAPMS.

Below is a brief description of each component of WAMPS:

- Reading Sensor: Generates a random value whose range is set based on the value of a "seriousness" variable.

- Reading Transmitter: Gets the generated value from the reading sensor and transmits it through the communicator.

- Power Controller: Each node will have a method called "turn on" that will start the node and we just call it. As for power-saving modes, this will depend on what the simulator will provide to us.

- Communicator: This is implemented by the simulator. Inter-Process communication is usually done using sockets; so, we expect the simulator to provide us with sockets as well as methods such as "send" and "receive".

- Launcher: Informs the data collector to start collection based on the delivery mode set by the user.

- Data Collector: Gets a list of nodes from which it has to collect readings, then sends messages to inform them and finally receives the required values.

- Aggregator: Implements the RCQ algorithm for data aggregation.

- Data Extractor: Use SQL queries to extract data from database.

- Data Displayer: This extracts data as required by the user and displays them in a table as well as evaluates the AQI for the selected area.

- Trend Analyser: Gets previous readings and determines relationship between them to be able to extrapolate future readings.

- Nodes Deployment Viewer: Displays deployment of nodes in the WSN field and their AQI colours.

- Connection Initiator: The java Driver Manager allows for a method to open a database, providing it the name of the database, user name and password as parameters. So, this component just has to make a call to this method and store the return reference to the connection.

- Connection Destructor: Connection object, in java.sql package, usually provides for a close method that closes the latter safely and frees associated memory as well as save the state of the latter. Therefore, this component just has to call this method.

The following table shows the various types of nodes that are present in WAPMS.

Table: Types of Nodes.

Type of Node	Energy Requirements	Location	Role
Source (sensor node)	Constrained	Random	Sensing and multihop routing
Cluster Head (collector)	Not Constrained	Fixed	Collection and aggregation
Sink /Gateway	Not Constrained	Fixed	Collection

These nodes will form a hierarchy that is shown in figure below:

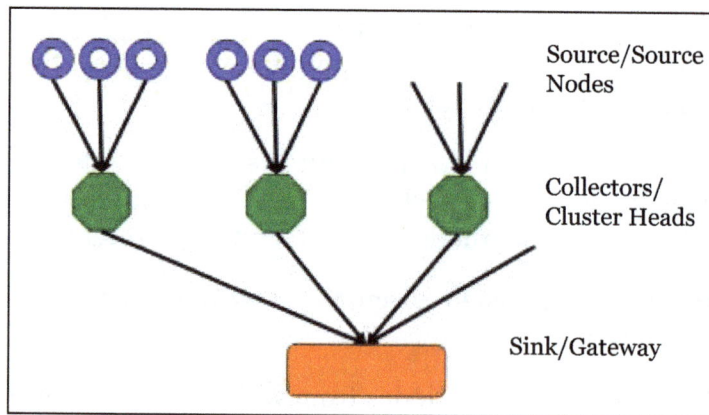

Hierarchy of nodes.

The strategy to deploy the WSN for our system is as follows:

- We first partition our region of interest into several smaller areas for better management of huge amount of data that will be collected from the system and for better coordination of the various components involved.

- We deploy one cluster head in each area; these will form cluster with the nodes in their respective areas, collect data from them, perform aggregation and send these back to the sink.

- We, then, randomly deploy the sensor nodes in the different areas. These will sense the data, send them to the cluster head in their respective area through multichip routing.

- We will use multiple sinks that will collect aggregated from the cluster heads and transmit them to the gateway. Each sink will be allocated a set of cluster heads.

- The gateway will collect results from the sinks and relay them to the database and eventually to our application.

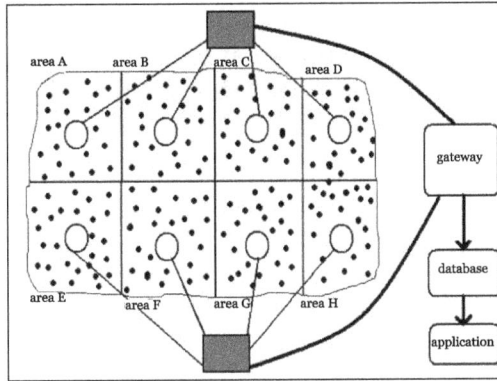

Deployment strategy.

The system is simulated over a small region as a prototype and then it will be extended to the whole island. The town of Port Louis, the capital of the country, is chosen for the prototype implementation as it is an urban area and therefore, more exposed to air pollution than rural areas. The site is partitioned the site into 6 smaller areas as shown in figure. With this small number of areas, we will use a single sink and we further simplify the system by allowing the gateway to play the role of the latter.

Partition of Port Louis into smaller areas.

An Air Quality Index (AQI) is used in WAMPS. The AQI is an indicator of air quality, based on air pollutants that have adverse effects on human health and the environment. The pollutants are ozone, fine particulate matter, nitrogen dioxide, carbon monoxide, sulphur dioxide and total reduced sulphur compounds.

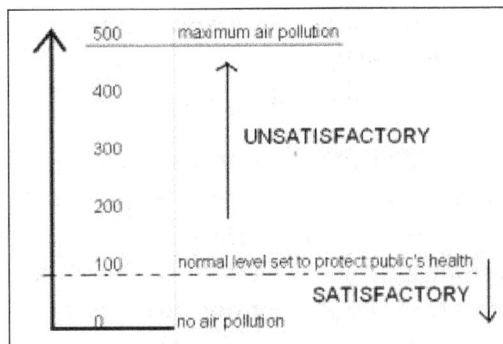

The range of AQI values.

The AQI consists of 6 categories, each represented by a specific colour and indicating a certain level of health concern to the public and is it shown in figure. The Ambient Air Quality Standards for Mauritius reports that the safe limit for ozone is 100 micrograms per m 3 and the safe AQI value set is also 100. Therefore, the AQI itself can, indirectly, be used to measure ozone concentration in Mauritius.

Table: Description of AQI categories.

Air Quality Index Levels of Health concern	Numerical Value	Meaning
Good	0.50	Air quality is considered satisfactory, and air pollution poses little or no risk.
Moderate	51-100	Air quality is acceptable; however, for some pollutants there may be a moderate health concern for a very small number of people who are unusually sensitive to air pollution.
Unhealthy For Sensitive Groups	101-150	Members of sensitive groups may experience health effects. The general public is not likely to be affected.
Unhealthy	151-200	Everyone may begin to experience health effects; members of sensitive groups may experience more seious health effects.
Very Unhealthy	201-300	Health alert: everyone may experience more serious health effects.
Hazardous	>300	Health warnings of emergency conditions. The entire population is more likely to be affected.

WAMPS has been simulated using the Jist/Swans simulator. JiST is a high-performance discrete event simulation engine that runs over a standard Java virtual machine. It converts an existing virtual machine into a simulation platform, by embedding simulation time semantics at the byte-code level. SWANS are a scalable wireless network simulator built atop the JiST platform. SWANS are organized as independent software components that can be composed to form complete wireless network or sensor network configurations. Its capabilities are similar to ns2 and GloMoSim but it is able to simulate much larger networks. SWANS leverages the JiST design to achieve high simulation throughput, save memory, and run standard Java network applications over simulated networks. In addition, SWANS implements a data structure, called hierarchical binning, for efficient computation of signal propagation.

The DSR protocol has been used for data transmission in WAPMS. The Dynamic Source Routing protocol is a simple reactive routing protocol designed specifically for use in multi-hop wireless ad hoc networks. DSR allows the network to be completely self-organizing and self-configuring, without the need for any existing network infrastructure or administration. DSR contains two phases: Route Discovery (find a path) and Route Maintenance (maintain a path). These only respond on a request. The protocol operates entirely on-demand, allowing the routing packet overhead of DSR to scale automatically to only that needed to react to changes in the routes currently in use.

After a collection, the system displays the nodes in their corresponding AQI colour as shown in figure. The following is an example of such a screen:

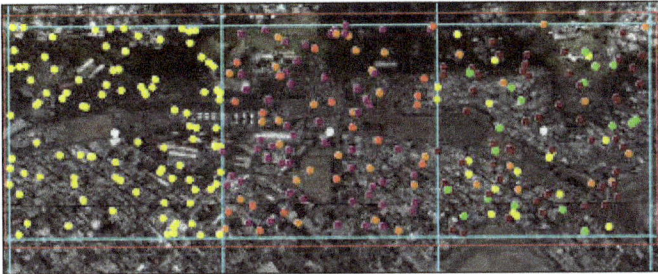

Nodes' Deployment after a collection.

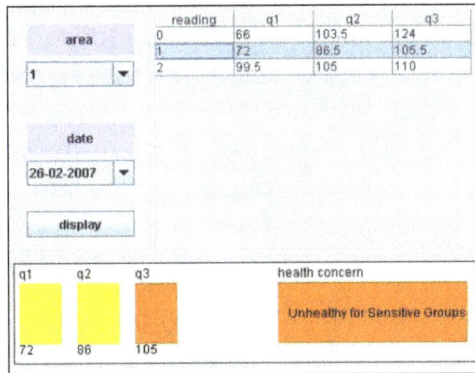

AQI for a selected area.

Furthermore, the WAPMS system allows fast analysis of received data through line graphs of selected areas as shown in figure.

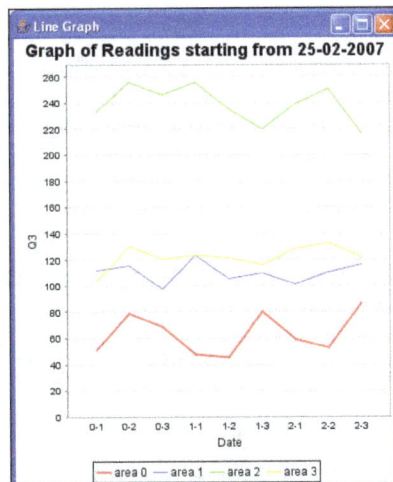

Line graph generated for selected areas.

The performance of WAPMS has been evaluation with increasing load. We have varied the number of areas simulated from 1 to 6 and for each case, we have varied the number of nodes per area from 50 to 200 and the execution time of the system has been recorded. The results are shown in table and figure.

Table: Execution Time of WAPMS.

Number of per area	Number of nodes per area			
	50	100	150	200
1	20	58	110	170
2	23	80	200	330
3	34	115	280	500
4	44	150	380	690
5	50	200	470	860
6	56	250	540	985

As shown in the above table, the maximum running time of our simulator is less than 20 minutes in the worst case of 6 areas and 200 nodes in each area. The short execution time of WAPMS is massively advantageous comparing to the existing air pollution monitoring unit of Mauritius that often takes days to measure pollution in an area. Moreover, WAPMS allows timely monitoring of an area and an abnormal situation can be detected almost immediately.

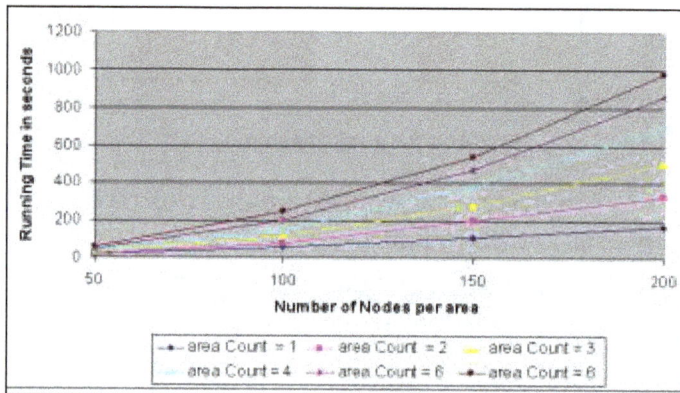

Performance analysis of WAPMS.

References

- Wireless-sensor-network, wireless-sensor-networks-technology-and-protocols: intechopen.com, Retrieved 30 June, 2019

- Soro S., Heinzelman W.: A Survey of Visual Sensor Networks, Advances in Multimedia, vol. 2009, Article ID 640386, 21 pages, 2009. doi:10.1155/2009/640386

- Sensor-Grid: senstar.com, Retrieved 2 May, 2019

- 43655919-A-Wireless-Sensor-Network-Air-Pollution-Monitoring-System: researchgate.net, Retrieved 22 August, 2019

Permissions

All chapters in this book are published with permission under the Creative Commons Attribution Share Alike License or equivalent. Every chapter published in this book has been scrutinized by our experts. Their significance has been extensively debated. The topics covered herein carry significant information for a comprehensive understanding. They may even be implemented as practical applications or may be referred to as a beginning point for further studies.

We would like to thank the editorial team for lending their expertise to make the book truly unique. They have played a crucial role in the development of this book. Without their invaluable contributions this book wouldn't have been possible. They have made vital efforts to compile up to date information on the varied aspects of this subject to make this book a valuable addition to the collection of many professionals and students.

This book was conceptualized with the vision of imparting up-to-date and integrated information in this field. To ensure the same, a matchless editorial board was set up. Every individual on the board went through rigorous rounds of assessment to prove their worth. After which they invested a large part of their time researching and compiling the most relevant data for our readers.

The editorial board has been involved in producing this book since its inception. They have spent rigorous hours researching and exploring the diverse topics which have resulted in the successful publishing of this book. They have passed on their knowledge of decades through this book. To expedite this challenging task, the publisher supported the team at every step. A small team of assistant editors was also appointed to further simplify the editing procedure and attain best results for the readers.

Apart from the editorial board, the designing team has also invested a significant amount of their time in understanding the subject and creating the most relevant covers. They scrutinized every image to scout for the most suitable representation of the subject and create an appropriate cover for the book.

The publishing team has been an ardent support to the editorial, designing and production team. Their endless efforts to recruit the best for this project, has resulted in the accomplishment of this book. They are a veteran in the field of academics and their pool of knowledge is as vast as their experience in printing. Their expertise and guidance has proved useful at every step. Their uncompromising quality standards have made this book an exceptional effort. Their encouragement from time to time has been an inspiration for everyone.

The publisher and the editorial board hope that this book will prove to be a valuable piece of knowledge for students, practitioners and scholars across the globe.

Index

www.ingramcontent.com/pod-product-compliance
Lightning Source LLC
Chambersburg PA
CBHW062005190326

41458CB00009B/2977

* 9 7 8 1 6 4 1 7 2 6 6 1 0 *